Carlin
October 1986

Concepts of Quantum Optics

Concepts of Quantum Optics

by

P. L. KNIGHT
Imperial College, London, UK

and

L. ALLEN
University of Sussex, UK

PERGAMON PRESS
OXFORD · NEW YORK · TORONTO · SYDNEY · FRANKFURT

U.K.	Pergamon Press Ltd., Headington Hill Hall, Oxford OX3 0BW, England
U.S.A.	Pergamon Press Inc., Maxwell House, Fairview Park, Elmsford, New York 10523, U.S.A.
CANADA	Pergamon Press Canada Ltd., Suite 104, 150 Consumers Road, Willowdale, Ontario M2J 1P9, Canada
AUSTRALIA	Pergamon Press (Aust.) Pty. Ltd., P.O. Box 544, Potts Point, N.S.W. 2011, Australia
FEDERAL REPUBLIC OF GERMANY	Pergamon Press GmbH, Hammerweg 6, D-6242 Kronberg-Taunus, Federal Republic of Germany

First edition 1983
Reprinted 1985

Library of Congress Cataloging in Publication Data
Knight, Peter (Peter L.)
Concepts of quantum optics.
Includes bibliographical references and index.
1. Quantum optics. I. Allen, L. (Leslie), 1935-
QC446.2.K6 1983 535'.15 83-8238

British Library Cataloguing in Publication Data
Knight, P. L.
Concepts of quantum optics.
1. Quantum optics
I. Title II. Allen L.
535.15 QC476.

ISBN 0-08-029150-3 (Hard cover)
ISBN 0-08-029160-0 (Flexicover)

Printed in Great Britain by A. Wheaton & Co. Ltd., Exeter

All these fifty years of conscious brooding have brought me no nearer to the answer to the question "What are light quanta ?" Nowadays every Tom, Dick and Harry thinks he knows it, but he is mistaken.

A. Einstein
1951

Preface

We have tried to write a book at a level accessible to the final year undergraduate and yet be appropriate and useful for a first year postgraduate student starting research in laser physics or quantum optics. It is not an attempt to write a definitive text book on quantum optics. The material discussed will need supplementation with more formal texts and reviews and we suggest some appropriate books at the end of this Preface.

Our intention is neither formality nor rigour but is an attempt to hint at the flavour of, and whet the appetite for, quantum optics. It is a guided tour of the field and at the same time an introduction to some of the original papers. We reprint here 17 papers, many of them well known; but again there has been no attempt to be definitive. Not every important paper is here, nor does every paper reproduced here report an original advance. It is, rather, a selection of papers which can be read by a keen but inexperienced student and which illuminates some of the points we make. We hope though that the idea of reproducing the papers will encourage the student to read primary sources and not to rely entirely on books. Although not intended to be comprehensive the book is, nevertheless, meant to be coherent and sequential. Our hope is that the six chapters will provide not just an introduction but some real insight into quantum optics.

The roots of the book lie in our continuing and long-standing interactions with other physicists who work in this area. Our contacts with Joe Eberly, Rodney Loudon, Leonard Mandel, Peter Milonni, David Pegg, Edwin Power, Carlos Stroud, Dan Walls and Emil Wolf have maintained and developed our enthusiasm for the subject. In some cases specific parts of the material discussed in this book have been learned at their feet and stolen from their hands. We wish, also, to thank Jean Hafner for typing drafts as well as the camera ready copy, Barbara Allen for assiduous proof reading and John Catherall for computer produced figures.

We are grateful in a quite different way to the authors of the papers reprinted in the latter part of this book who readily gave their consent. We are pleased to thank and acknowledge their publishers, Cambridge University Press (Paper 1), Taylor and Francis (Paper 3), The American Physical Society (Papers 4, 8 and 16), The American Journal of Physics (Papers 5, 10, 17), North-Holland Publishing Co. (Paper 6), "Nature" (Papers 7, 13, 14 and 15) and the Institute of Physics (Papers 9, 11, 12) for giving permission to reproduce the papers.

The book needs to be read side by side with the appropriate parts of other more formal books. In particular we suggest:

Allen, L. and Eberly, J.H.
"Optical Resonance and Two-Level Atoms"
Wiley, New York, 1975.

Haken, H.
"Light" Vol. I
North Holland, Amsterdam, 1981.

Loudon, R.
"The Quantum Theory of Light"
O.U.P., 1973.

Nussenzveig, H.M.
"Introduction to Quantum Optics"
Gordon and Breach, New York, 1973.

Sargent, M., Scully, M.O. and Lamb, W.E. Jr.
"Laser Physics"
Addison-Wesley, Reading, 1973.

while the following

Mandel, L. and Wolf, E.
"Selected papers on Coherence and Fluctuations of Light"
Vols. I and II.
Dover, 1970.

Schwinger, J.
"Quantum Electrodynamics"
Dover, 1958.

provide invaluable collections of seminal papers in this field.

We hope the book will be a useful adjunct to the more formal texts; we learned a great deal writing it and hope others may gain too. Our only fear is that Einstein would have thought our names were Tom, Dick or possibly Harry.

P.L. Knight
L. Allen

October 1982

Contents

CHAPTER 1

Radiation and Quanta

1.1 THE EARLY HISTORY OF RADIATION

As a subject of continuing interest light stands the test of time very well. It is believed by many to have appeared very soon after the beginning and in a corpuscular form as "God divided the light from the darkness" (Genesis 1.4), which appears to eliminate the phenomenon of diffraction. The competing notions of wave and particle have dominated our thinking about light ever since. Useful discussions of the history of our ideas of radiation have been given by Whittaker (1951). Empedocles of Agrigentum was writing in the fifth century B.C. of a corpuscular theory of light moving at a finite speed. Euclid in 300 B.C. had formulated the law of reflection, and Seneca, who died in 65 A.D., produced colours by passing white light through a prism. The Platonic tactile notion of vision was widely adopted following the proposition by Pythagorus. This theory considered that a stream of particles emitted by the eye travelled in straight lines to the object under scrutiny, combined with solar rays at the object and returned to the eye to give sight. An alternative idea, recorded by Epicurus about 300 B.C., on emission theory, argued that the object emitted the light and created sight as it entered the viewer's eye. The tactile theory was not finally rejected until the work of Ibn al-Haythen, or Alhazon, who knew enough about reflection of the emitted light to put the angles of incidence and reflection in the same plane, normal to the reflecting surface.

From the early seventeenth century to the late nineteenth, argument as to the nature of light rays was vigorous. Grimaldi discovered diffraction in 1665 and in the same year Hooke suggested light was a transverse wave phenomenon. Newton, however, laid stress on the rectilinear propagation of light rays, and, realising that physical objects would travel in straight lines in the absence of any applied forces, proposed a corpuscular theory of light. Although diffraction had already been observed and although he himself investigated interference effects, Newton believed that strict rectilinear propagation was not compatible with wave theory because waves spread out in all directions. The observation of polarization by Grimaldi and by Bartholinus was also recognised as incompatible with a longitudinal wave but Huygens realised that a transverse wave would allow such phenomena. On balance, Newton felt that the corpuscular theory was needed but he never completely rejected wave theory. He thought that the corpuscles had "fits of easy reflection" and "fits of easy transmission" to account for the interaction of light at a surface. In his book "Opticks", Newton considered space to be permeated by an elastic medium called the aether, capable of propagating waves of high velocity but did not identify these waves with light. He considered light to be emitted by sources, probably in corpuscular form, but in interaction with the aether waves. Reflection, refraction and diffraction he thought to be due to a coupling of light with the aether at the surfaces.

The transverse wave theory of light due initially to Huygens, and supported by de Malebranche who in 1699 proposed that the colour of light was proportional to the wave period and its brightness to the wave amplitude, was greatly developed by Thomas Young at the beginning of the nineteenth century. Young used the transverse wave theory to explain interference and the colour of thin films, and this was developed by Fresnel, Kirchhoff and others. This concept required an all-pervasive medium or aether, with the properties of an elastic solid, within which the light travels.

1.2 MAXWELL'S ELECTROMAGNETIC THEORY OF LIGHT

In 1865, James Clerk Maxwell published his *A Dynamical Theory of the Electromagnetic Field* and connected the electric field $\underline{E}$ and the magnetic field $\underline{B}$ with the charge density ρ and the current density $\underline{j}$, incorporating Faraday's empirical observations such as induction. Maxwell's equations for $\underline{E}$ and $\underline{B}$ unified our understanding of the two previously separate forces of electricity and magnetism. He deduced the electromagnetic wave equation which predicted the propagation of electromagnetic disturbances in a mechanical aether at the experimentally measured speed of light. His theoretical work was difficult to read, and it took twenty years before Hertz generated electromagnetic waves of a frequency ~ 75 MHz and detected them in the laboratory. Heaviside and Poynting in 1885 showed that electromagnetic fields carried energy and momentum. Classical electrodynamics developed so rapidly in the late nineteenth century that Lodge was able to say in 1888 that "the whole subject of electromagnetic radiation seems working out splendidly."

While the classical theory of electromagnetism was achieving such success, a few troublesome disagreements with experiments in radiative interactions began to appear. Experiments by Lenard, when attempting to explain earlier observations by Hertz, showed that negatively charged metal lost its charge when exposed to uv radiation. That is, the electrons gained energy from the radiation and were able to escape from the metal. It was found that the number of emitted electrons increased with increasing light intensity but that the kinetic energy of the electrons only increases with increasing frequency of the incident light. It was this observation which presented the difficulty for classical theory, which required the kinetic energy of the electrons to depend simply on the intensity of the incident radiation and not on frequency.

Classical electrodynamics proved to be similarly inadequate to account for the spectral distribution of light emitted by a black-body at temperature T. Stefan had found empirically that the rate of heat loss was proportional to T^4 and this had been deduced by Boltzmann from Maxwell's theory and thermodynamics. Wien, a few years later, was able to predict the entire emission curve at any temperature from his displacement law if the entire curve was known at any other temperature. He was able to show that the wavelength of the maximum intensity of emission was proportional to 1/T. Yet there was no theory which accounted for the frequency distribution of the emission. Rayleigh, and to some extent, Jeans, using classical theory achieved agreement with experiment at low frequencies but great disagreement with the behaviour at high frequencies. The essence of the theory was that the law of equipartition of energies should apply to the electromagnetic field in an enclosure. A standing wave of a particular frequency interacts with oscillating atoms in the wall allowing energy to flow either from wall to field, or field to wall. Rayleigh calculated how many different standing waves, or modes of radiation there were within a certain interval of wavelength and, although he assumed that the energy in a mode could have any value, attributed an average energy kT to each mode. The full classical derivation may be read in any number of quantum mechanics or atomic physics books and will not be repeated here. The classical Rayleigh-Jeans law predicts that the energy per unit frequency interval, $E(\nu)d\nu$ is determined by

$$E(\nu)d\nu = \frac{8\pi\nu^2 kT}{c^3}\,d\nu \tag{1.1}$$

and is unbounded as ν increases. The total emission energy

$$E = \int_0^\infty E(\nu)d\nu$$

is ultraviolet-divergent. In other words classical electrodynamics predicts that all the energy in the Universe would be carried by high frequency waves.

1.3 PHOTONS AND THE OLD QUANTUM THEORY

Even while Rayleigh and others were working on black-body radiation in classical electrodynamics, Planck had devised an equation which agreed with the whole of the experimentally determined black-body emission spectrum. He did this initially by changing constants in Wien's analysis. To justify this he was obliged to assume that the walls of the black-body gave off, or absorbed, discrete quanta or packets of energy. This was the beginning of what we now call the old quantum theory (ter Haar 1967, Pais 1979). Planck assumed that the walls were composed of oscillators in thermal equilibrium which possess discrete energy levels and not the continuous range associated with a classical oscillator. An oscillator of frequency ν was assumed to be only capable of sustaining the energies $nh\nu$ where n is an integer. Planck then argued that each field mode could only receive energy from the wall oscillators in discrete amounts, and derived his black-body energy distribution formula

$$E(\nu)d\nu = \frac{8\pi^2\nu^2}{c^3}\;\frac{h\nu}{(\exp(h\nu/kT) - 1)}\,d\nu \tag{1.2}$$

where h is Planck's constant. In the long wavelength limit, $\exp(h\nu/kT) \rightarrow (h\nu/kT) + 1$ and eq.(1.2) reduces to the Rayleigh-Jeans law eq.(1.1). If eq.(1.2) is integrated over all frequencies we obtain Stefan's law

$$E = \int_0^\infty E(\nu)d\nu = \frac{8}{15}\,\frac{\pi^5}{c^3h^3}\,(kT)^4 \tag{1.3}$$

Planck quantized the wall oscillators but did not consider the electromagnetic radiation to be quantized.

Einstein's paper of 1905 contains not only his Nobel prize winning discussion of the photoelectric effect, but a lengthy discussion of black-body radiation in which he showed that the Planck formula suggests light behaves as a gas of corpuscles (ter Haar 1967, Milonni 1982). Einstein showed that the entropy of black-body radiation can be written as

$$S = V \int_0^\infty d\nu \; f(\rho,\nu) \tag{1.4}$$

where V is the cavity volume. He assumed that different frequency components are independent and that the radiation is fully characterised by $\rho(\nu)$. The function f is governed by the equation

$$\frac{\partial f}{\partial \rho} = \frac{1}{T} \tag{1.5}$$

which he derived by maximising the entropy for a given energy. From this and the Wien law, valid for high frequencies,

$$\rho(\nu) = \alpha\nu^3 \exp(-\beta\nu/T) \tag{1.6}$$

where β is Wien's constant, we find for the entropy

$$S = -\frac{V}{\beta}\int_0^\infty d\nu \; \frac{\rho(\nu)}{\nu}\left[\log\left(\frac{\rho(\nu)}{\alpha\nu^3}\right) - 1\right] . \tag{1.7}$$

But the energy in the frequency interval $d\nu$ is

$$E(\nu)d\nu = V\rho(\nu)d\nu \tag{1.8}$$

so from eq. (1.7) we find the entropy

$$S(\nu) = -\frac{E(\nu)}{\beta\nu}\left[\log\left(\frac{E(\nu)}{V\alpha\nu^3}\right) - 1\right] . \tag{1.9}$$

If the energy is held fixed while changing the volume of the cavity from V_1 to V_2 reversibly the entropy change, from eq. (1.9), is

$$S_2(\nu) - S_1(\nu) = \frac{E(\nu)}{\beta\nu}\log\left(\frac{V_2}{V_1}\right) . \tag{1.10}$$

Einstein compared eq. (1.10) to the entropy change of an ideal gas whose volume is likewise isothermally changed from V_1 to V_2,

$$S_2(g) - S_1(g) = Nk \log\left(\frac{V_2}{V_1}\right) \tag{1.11}$$

where N is the number of particles in an ideal gas. These expressions are identical if $E(\nu) = \beta\nu Nk$. As the Wien law agrees with the Planck law for high frequencies if $\beta = h/k$,

$$E(\nu) = Nh\nu \quad . \tag{1.12}$$

If the radiation satisfies Wien's law, that is the high frequency part of the Planck law, then it behaves as if it were composed of independent quanta of energy $h\nu$. Einstein was careful to point out that non-Wien radiation, for example at low frequencies, would not behave in this way. Indeed if $h\nu/kT \ll 1$, the Rayleigh-Jeans limit, Einstein later found a classical wave interpretation of the thermodynamic properties of black-body radiation.

Einstein went on in his 1905 paper to provide a full theoretical description of the photoelectric effect. Using the corpuscular picture suggested by eq.(1.12) Einstein showed that conservation of energy leads to an equation for the kinetic energy of the emitted photoelectron

$$\frac{1}{2}mv^2 = h\nu - W \ , \tag{1.13}$$

where the work function W is the minimum energy required to liberate the electron from the metal and depends on the metal used. Einstein's result showed that the velocity of the emitted electrons would be independent of the incident light intensity while the number emitted would be proportional to the intensity. Millikan verified eq.(1.13) and from the slope of his graph of the kinetic energy of the electrons against the frequency of the emitted light, found the value of h to agree with that chosen by Planck. The fact that the photoelectric effect can now be described using semiclassical theory does not, of course, diminish its importance for the development of quantum ideas. Einstein also successfully applied Planck's idea of quantized oscillators to the theory of the specific heat of solids, where classical theory was known to be wrong at low temperatures.

The revival of corpuscular ideas of light in the first decade of the twentieth century, although unsupported by a real theory, lead to the first attempts to find quantum effects in coherence experiments. G.I. Taylor in 1909, following a suggestion from J.J. Thomson, performed an experiment to see whether for extremely weak light sources the indivisible quanta would modify ordinary diffraction effects (Taylor, Reprinted Paper 1). He used such dim light that in modern language we would say that there was never more than one photon arriving at his photographic plate at a given time. His exposure time was about three months. Taylor observed a diffraction pattern of the same form regardless of the mean intensity, with the fringes built up by successive quanta.

In 1909, Einstein returned to the problem of the thermodynamic properties of radiation and provided the first clear evidence of wave-particle duality (Einstein 1909). He abandoned Wien's formula and used instead Planck's formula for the energy density $\rho(\nu)$. He had previously shown that the energy fluctuations of a system in thermal equilibrium are described (Pais 1979) by,

$$\langle E^2\rangle - \langle E\rangle^2 \equiv \langle(\Delta E)^2\rangle = kT^2 \frac{\partial}{\partial T} \langle E\rangle \ . \tag{1.14}$$

For thermal radiation of frequency ν, we know that $\langle E(\nu)\rangle = V\rho(\nu)$. From eq. (1.14) and the Planck law for $\rho(\nu)$ Einstein obtained the fluctuation formula

$$\langle(\Delta E(\nu))^2\rangle = h\nu\rho(\nu) + \left[\frac{c^3}{8\pi\nu^2} \ \rho^2(\nu)\right] V \tag{1.15}$$

where the energy density is given by the Planck form

$$\rho(\nu) = \frac{8\pi h\nu^3 / c^3}{(\exp(h\nu/kT) - 1)} \ . \tag{1.16}$$

If the Rayleigh-Jeans formula

$$\rho_{RJ}(\nu) = \frac{8\pi\nu^2}{c^3} kT \tag{1.17}$$

had been used to derive the fluctuation in energies of classical light waves, $<(\Delta E(\nu))^2>_W$ using eq.(1.14), we would find

$$<(\Delta E(\nu))^2>_W = \frac{c^3}{8\pi\nu^2} \rho^2(\nu)V \quad . \tag{1.18}$$

The Wien law (eq.(1.6)), valid at high frequencies, leads to a particle-like energy fluctuation, $<(\Delta E(\nu))^2>_p$,

$$<(\Delta E(\nu))^2>_p = h\nu\rho(\nu)V = h\nu E(\nu) \quad . \tag{1.19}$$

This result, eq.(1.19), is characteristic of the classical Poissonian statistics of distinguishable particles. We see that the general Einstein fluctuation formula eq.(1.15) can be written as

$$<(\Delta E(\nu))^2> = <(\Delta E(\nu))^2>_p + <(\Delta E(\nu))^2>_W \tag{1.20}$$

showing that black-body radiation exhibits fluctuation properties characteristic of both particles and waves.

We can write the Einstein fluctuation formula eq.(1.15), using eq.(1.16), as

$$<(\Delta E(\nu))^2> = h\nu\rho(\nu) \left[1 + \frac{1}{\exp(h\nu/kT) - 1}\right] V \quad . \tag{1.21}$$

In the high-frequency Wien limit $h\nu/kT \gg 1$

$$<(\Delta E(\nu))^2> \approx h\nu\rho(\nu)V = <(\Delta E(\nu))^2>_p \tag{1.22}$$

and we see that black-body radiation behaves apparently as a gas of independent particles. In the low-frequency, or Rayleigh-Jeans limit, $h\nu/kT \ll 1$

$$<(\Delta E(\nu))^2> = <(\Delta E(\nu))^2>_W \tag{1.23}$$

the black-body radiation behaves as a superposition of classical waves. The classical limit when $h \to 0$ gives only the wave fluctuations and the simultaneous presence of wave and particle properties in eq.(1.15) is a quantum effect requiring a non-vanishing Planck's constant.

The next vital component in the development of the old quantum theory was due to Bohr. He combined Planck's quantum ideas with the concept of Rutherford's 'nuclear atom'. Rutherford believed from scattering experiments that an atom consisted of a small positively charged nucleus surrounded by electrons. However the Rutherford model neither accounted for the wealth of observed atomic spectra already accumulated, nor was it possible to devise a stationary array of charged particles which could remain in stable equilibrium under their own electrostatic forces. Thus the electrons had to be in motion and their acceleration as they orbited the nucleus would result in electromagnetic radiation causing energy loss and thus necessitating their spiral into the nucleus. Apart from the atom no longer being stable, the spectrum emitted would be continuous and not consist of the discrete spectral lines characteristic of each element.

Bohr assumed that the electron could only revolve about the nucleus in a stationary state of energy in which no electromagnetic radiation was emitted and that these states, or orbits, were those in which the angular momentum of the electron about the nucleus was $nh/2\pi$, where n is again an integer. This assumption, of course, again violates classical theory because an electron in such an orbit would still radiate. However, in spite of that, his second assumption was that emission or absorption of radiation only occurred when the electron made a transition, or quantum jump, from one energy state, E_i, to another, E_f. The energy of the emitted quantum is given by

$$E_i - E_f = h\nu \quad . \tag{1.24}$$

Sommerfeld showed that the Bohr quantum condition could be expressed as

$$\oint p\,dq = nh \tag{1.25}$$

where p is the momentum conjugate to the position q and the integration is over one orbit. He developed Bohr's theory to include elliptic as well as circular orbits when the above condition had to be generalized to two, and involve angular and radial quantum numbers. More importantly, when the Sommerfeld approach is applied to a linear harmonic oscillator the energy levels obtained are those hypothesised by Planck. In the space of less than 20 years great advances had been made although there still remained uncomfortable hybrids of quantum and classical ideas. However the Bohr-Sommerfeld model, for all its success, was still in conflict with classical theory.

In 1917 Einstein studied the properties of an ensemble of atoms in thermal equilibrium with black-body radiation. To allow the radiation to be described by the Planck formula he found it necessary to postulate the existence of both spontaneous and stimulated transitions. (See Reprinted Paper 2).

Compton developed the idea of energy quanta further in his consideration of the scattering of x-rays for which von Laue had already established a wave character by diffraction experiments. Compton found that when short wavelength x-rays were scattered by light mass atoms the scattered radiation included x-rays of longer wavelength as well as that of the incident x-ray. Classical theory, which accounts for Thomson scattering, requires the scattering atomic electrons to vibrate with the frequency of the incident radiation and so emit only the same frequency. However Compton assumed that the radiation consisted of Einstein's quanta of energy $h\nu$, each with momentum $h\nu/c$ in agreement with relativistic particle kinematics. Combining this idea with conservation of energy and momentum and the relativistic kinetic energy of the electron in recoil he found, in reasonable agreement with observation, that

$$\Delta\lambda = \lambda_{sc} - \lambda_{inc} = (h/m_o c)(1-\cos\theta) \tag{1.26}$$

where λ_{inc} and λ_{sc} are the incident and scattered wavelengths respectively, θ the scattering angle and m_o the rest mass of the electron. Compton scattering arises when an electron is knocked out of the atom and so absorbs energy and momentum. The energy for this can only come from a decrease in energy of the incident light quanta which increases it's wavelength. The word "photon", but not its currently accepted implications, was introduced in 1926 by G.N. Lewis and has tended to supercede the cumbersome phrase "light quantum".

The idea of radiation as quanta, now substantiated by Einstein and by Compton whose work showed the quanta carried well directed momenta as well as energy seemed to be in sharp conflict with the well established wave representation of interference and diffraction. The idea that it was not just light but all matter which possessed this dual nature was due to de Broglie. He suggested that any moving particle has a wavelength associated with it such that

$$p = h/\lambda \qquad (1.27$$

which, of course, is consistent with Compton's value of the momentum of the radiation quantum $h\nu/c$. The proposed waves were not electromagnetic waves but matter waves which played the role of guiding the particle. The Bohr-Sommerfeld condition for a circular atomic electron orbit with angular momentum p is

$$\oint p\, dr = nh \qquad (1.28)$$

which gives

$$(2\pi r)p = nh\ . \qquad (1.29)$$

If p is h/λ, then the circumference of the circular orbit, $(2\pi r)$, is equal to $n\lambda$, or an integral number of de Broglie wavelengths: this suggests that atomic particles possess wavelike properties and that electrons in stationary states are represented by standing waves.

The nature of the duality to some extent presents no problems. It is possible to demonstrate for example that the trajectories of particles are entirely compatible with wave theory. But a wave incident at an interface in general splits into reflected and refracted components. The concept of a divided particle is not so enticing, especially if the particle is charged. However de Broglie suggested that the particles are "real" and that the waves guide the particles. The modern interpretation due to Born is that the waves may be considered as probability waves where the amplitude of the waves at a point is a measure of the probability of the particle being there. Within this interpretation, the amplitude of a reflected wave determines the probability of the particle being reflected at the interface. That is, a single particle will either reflect or refract but a large number of particles will divide in accordance with the probability. We have to take the average of a large number of events to describe the relative probabilities of the two processes. Nevertheless the problem of photons in a two-slit experiment [See the paper by Frisch, Reprinted Paper 3] is not trivial even within this interpretation. The dark and bright fringes represent regions of high and low probability for the arrival of the photon. If one slit is covered the interference pattern changes and where there was originally zero probability of the photon arriving, there is now a finite probability. An individual photon must go through one slit or the other, yet its subsequent route in space is affected by the other slit which may or may not be open; yet, how could it "know" not having been through that slit whether or not it was open? The difficulty arises when the two slits are open when we write as if we knew through which slit the particle passed. We do not know: if we did there would be no interference pattern. The act of detecting the photon would destroy the interference. We are not able to detect both the particle and the wave nature of our particles simultaneously. In contrast neither the wave model nor the particle model is sufficient by itself; wave-particle duality is a necessary synthesis.

1.4 QUANTUM MECHANICS

Arising from the debates and developments briefly outlined above, the years 1925 and 1926 saw the origin of several different strands of our contemporary view of quantum mechanics.

Heisenberg introduced matrix mechanics and found it necessary to employ non-commutative algebra. Born realised that this approach was precisely that of the calculus of matrices discovered by Cayley and together with Jordan found that the commutator of p and q involved Planck's constant, namely $pq-qp = -i\hbar$. A few months later, in 1926, Pauli applied the theory to the hydrogen atom and at the same time Dirac produced a more abstract, although ultimately more elegant theory and applied it to the same atom. In so doing he demonstrated the close relationship between classical and quantum mechanics by connecting the quantum mechanical commutators with the classical Poisson brackets. In the same year Schrödinger published the results of his work on wave mechanics and where Bohr had found it necessary to postulate the need for quantization and quantum numbers, Schrödinger derived them using his wave equation for the evolution of a complex wave-amplitude ψ which in some sense represented de Broglie's matter waves. We shall not discuss the derivation of wave mechanics in detail nor even that of the time-independent wave-equation

$$\frac{\hbar^2}{2m}\nabla^2\psi + (E-V)\psi = 0 \tag{1.30}$$

and the time-dependent one

$$-\frac{\hbar^2}{2m}\nabla^2\psi + V\psi = i\hbar\frac{\partial\psi}{\partial t} , \tag{1.31}$$

suffice to say they were developed from classical wave ideas. But, despite the difference in the approach used, Schrödinger in 1926 demonstrated the equivalence of matrix and wave mechanics by achieving the same result as Born and Jordan for the commutator $[p,q]$.

Bose (1924) presented a derivation of Planck's black-body radiation spectrum using quantum statistics by counting cells in phase-space. A little later Einstein (1924, 1925) derived an equivalent expression by counting the number of ways N indistinguishable photons can be distributed among Z cells of phase space. For such "Bose-Einstein" particles the mean number of particles in state j is

$$\langle n_j\rangle = [\exp(E_j/kT) - 1]^{-1} . \tag{1.32}$$

If we assume that the smallest region of phase space has size h^3, then the number of such cells, Z_p, in the phase space of a particle in a volume V with momentum in the interval from p to p + dp is

$$Z_p = \frac{1}{h^3}\int_V d^3x\ d^3p = \frac{4\pi}{h^3} V\ p^2 dp \quad . \tag{1.33}$$

Photons have momentum $h\nu/c$, so

$$Z_p = \frac{4\pi}{c^3} V\ \nu^2 d\nu \quad . \tag{1.34}$$

This has to be multiplied by two to account for two independent polarizations and then equals the number of states, or modes, of the field in the frequency interval from ν to $(\nu + d\nu)$. The energy density, from eqs.(1.32) and (1.34) is

$$\rho(\nu) = \frac{8\pi\nu^2}{c^3} (h\nu) \langle n_\nu \rangle \tag{1.35}$$

This enables us to write Einstein's fluctuation formula eq.(1.15) in terms of the photon number fluctuations

$$\langle \Delta n_\nu^2 \rangle = \langle n_\nu \rangle + \langle n_\nu \rangle^2 \tag{1.36}$$

and provides a link between indistinguishability, Bose-Einstein counting and wave fluctuations. A further ingredient of the quantum description of wave-particle duality is Heisenberg's Uncertainty Principle formulated in 1927. This principle gives a quantitative estimate of the uncertainty in the position Δx of a wave-like object of uncertain momentum Δp, which arises because waves cannot be localized in space.

$$\Delta p \Delta x \sim \hbar \ .$$

Bohr's principle of complementarity says the same thing in another way. If an experiment allows one aspect of a physical phenomenon to be observed it simultaneously prevents us from observing a complementary aspect of the same phenomenon. Thus if we know through which slit the photon passes (see Frisch, Reprinted Paper 3) we do not observe its wave-like property. Closing one slit does not yield an interference pattern; the fact of recording an interference pattern necessarily makes the determination of which slit the photon passed through impossible. It should be realised that the uncertainty principle is not a trick designed to get around the problems of wave-particle duality. It may be logically deduced and no experiment is known which violates it; it is entirely compatible with, and contained within, quantum mechanics. This chapter is not meant in any way to be a substitute for a real historical review of quantum theory. An excellent account of this is to be found in the book by Jammer (1966). One aspect of the theory, however, remains to be elaborated upon. This is the concept of probability amplitude and its relationship with the ideas of wave-particle duality.

In Schrödinger's wave equation the wave function $\psi(r,t)$ describes the motion of a particle in space. The currently accepted interpretation of ψ is that the probability density of finding the particle at a given point in space is expressed by $|\psi|^2$. That is,

$$P(r,t)d^3r\,dt = |\psi|^2 d^3r\,dt \quad . \tag{1.37}$$

This is the probability of finding the particle at time t within the volume element d^3r. As the particle must be somewhere, we can normalize ψ by recognizing that

$$\int |\psi|^2 d^3r = 1 \tag{1.38}$$

In much of quantum physics it is found that a system possesses not one but a number of eigenstates and that, in general, the wavefunction describing the system may be written as

$$\psi = \sum a_i \psi_i \ . \tag{1.39}$$

By multiplying each side of eq.(1.39) by ψ_i^* and using the orthogonality condition for the wavefunctions

$$a_i = \int \psi_i^* \ \psi dr^3 \ , \tag{1.40}$$

or to use the Dirac notation

$$a_i = \langle \psi_i | \psi \rangle \tag{1.41}$$

we see that the normalization may be written as

$$\int |\psi|^2 dr^3 = \int \psi^* \psi dr^3 = \sum_i |a_i|^2 = 1 \quad . \tag{1.42}$$

It follows that $|a_i|^2$ is the probability that a measurement on the system will find it in the eigenstates described by the wavefunction ψ_i. The coefficient itself, a_i, is usually called the probability amplitude.

When there are two indistinguishable ways for a particle to reach a final position in space it is found that the resulting probability is not the sum of the two probabilities but the absolute square of the sum of the two amplitudes. We know of no better discussion of this point than that of Feynman and strongly recommend reading Chapter 37 in Vol.I and Chapter 3 in Vol.II of "The Feynman Lectures in Physics". If we borrow shamelessly from him and consider a light quantum going from a source s to either slit 1 or to slit 2 in a Young's two-slit experiment, and from either slit to an observation point x it follows that the amplitude for the process may be represented by,

$$\langle x|s\rangle = \langle x|1\rangle\langle 1|s\rangle + \langle x|2\rangle\langle 2|s\rangle \tag{1.43}$$

and the probability for the process will be given by the square modulus of this expression.

We may write rather more generally,

$$\langle x|s\rangle = \sum_i \langle x|i\rangle\langle i|s\rangle \qquad (1.44$$

and

$$|\langle x|s\rangle|^2 = |\sum_i \langle x|i\rangle\langle i|s\rangle|^2 \quad . \qquad (1.4$$

Clearly $<x|s>$ is equivalent to our probability amplitude, a_i, as given in eq.(1.42). Equally clearly if we address ourselves to the problem of two slit interference once again, the value of $|<x|s>|^2$ contains cross-terms involving the routes 1 and 2 and this represents the interference. In other words the matter waves associated with the two routes interfere just as in the superposition of waves required to explain Young's experiment classically. As Feynman eloquently discusses, the amplitudes for different and distinguishable final states must not be added. We must square the amplitudes of all possible final events and then sum. If only one slit or the other was available at any time, or if the route chosen by a specific light quanta was determinable then,

$$|<x|s>|^2 = |<x|1><1|s>|^2 + |<x|2><2|s>|^2 \quad (1.46)$$

and the interference term vanishes.

1.5 DIRAC'S RADIATION THEORY

The radiation field is dynamically equivalent to an infinite set of uncoupled harmonic oscillators, one for each mode of the electromagnetic field. Born, Heisenberg and Jordan (1926) used this harmonic oscillator decomposition to quantize the radiation field according to matrix mechanics. One product of their analysis was a quantum field theoretical derivation of the Einstein fluctuation formula including both wave and particle contributions. This fully quantized derivation firmly links the particle contribution in the fluctuation formula with field commutators and the occurrence of a field zero-point energy. Dirac (1927) showed how to systematically quantize the free field as an independent dynamical system, again by exploiting the representation of each field mode as a harmonic oscillator. Dirac associated combinations of oscillator momentum and position operators called annihilation and creation operators with the mode amplitudes. The single excitation of a mode, or state, of the radiation field represents the presence of a photon in that mode. In the context of Young's two-slit experiment, the photon is a quantized excitation of one of the normal modes of the entire system. It is with this understanding that Dirac's famous comment that "the photon interferes only with itself" must be viewed. In Chapter 2 we will examine in detail the photon picture of interference and correlation and qualify Dirac's remark. This modern idea of a photon is a long way from the old quantum theory corpuscular, localized "photon", and bears no relationship to "fuzzy ball" wave packets invoked in many popular elementary texts (see Lamb 1973).

Dirac also described how the quantized electromagnetic field interacts with matter, and accounted for the phenomena of stimulated and spontaneous transitions in atoms. Since in quantum mechanics an oscillator can never be at rest, except at the expense of an infinite momentum, each mode of the radiation possesses zero-point oscillation in the ground, or vacuum state. The vacuum in quantum electrodynamics is visualized as a ferment of activity. Such vacuum fluctuations can couple to the motion of electrons in atoms and change the energy of atomic states (see Reprinted Papers by Lamb and Retherford, 4, and by Power, 5).

In what follows we will discuss how the radiation field is quantized; the modern view of a photon as the occupation of a cavity field mode; and the problems and scope of quantum optics. It is sensible at this stage not to dwell on formalism. Rather we now wish to pose a question: "When do we need the photon in the physics of the interaction of radiation and matter?" We will not concern ourselves with philosophy or with history: instead we shall attempt to answer the question and, with the answer in mind, develop just enough formalism to enable us to study the appropriate parts of atomic physics. The truth is that, whatever idea of optical radiation we may have at the back of our minds, a great many interactions of light and matter may be solved without invoking the photon at all. We find that semi-classical theory, where the radiation is represented by a classical wave and the atom is quantized, is fully sufficient for a great many purposes. In view of the preceding discussion such a result will have to be justified.

We shall find two main areas where the photon behaviour of light needs explicitly to be taken into account. First, when spontaneous emission is involved and second for the understanding of intensity correlations in or between light beams, or where the behaviour of an atom is dictated by the intensity correlation of a source of radiation.

Spontaneous emission involves more than just the incoherent relaxation of an excited state, it includes for example the Lamb shift which is the result of the same interactions which give rise to spontaneous emission. The observation of intensity correlations appeared to re-open philosophical debate about photons in the 1950's, although an understanding was soon achieved without the need to create any new concept. It is though pleasing, as we shall see in the final Chapter of this book, that new phenomena continually arise which call upon the different parts of our understanding and force us to seek new explanations. Photon anti-bunching is simply the newest in a long line of radiation phenomena.

CHAPTER 2

Quantization of the Radiation Field

2.1 CAVITY MODES

In this Chapter we present an outline of the description of quantized radiation fields. Full discussion of the topic may be found in the textbooks by Power (1964), Loudon (1973) and Sargent, Scully and Lamb (1974). We first discuss the properties of the radiation field in a bounded cavity, show how each cavity field mode has the characteristics of a simple harmonic oscillator and how each can be quantized using known results for the quantum harmonic oscillator. We then discuss the quantum fluctuations of the radiation field and the interpretation of a photon as an occupation of a normal mode of the system.

We consider a radiation field confined in a one-dimensional cavity consisting of two perfectly conducting walls with no sources of radiation, either as currents or charges, present in the cavity. We imagine the free field to be described by an electric field vector $\underline{E}(\underline{r},t)$ polarised in the x-direction. Maxwell's equations for the source-free radiation are

$$\underline{\nabla} \times \underline{H} = \frac{\partial \underline{D}}{\partial t} \tag{2.1}$$

$$\underline{\nabla} \times \underline{E} = -\frac{\partial \underline{B}}{\partial t} \tag{2.2}$$

$$\underline{\nabla}.\underline{B} = 0 \tag{2.3}$$

$$\underline{\nabla}.\underline{E} = 0 \ , \tag{2.4}$$

where $\underline{B} = \mu_0 \underline{H}$ and $\underline{D} = \varepsilon_0 \underline{E}$. The electric field satisfies a periodic boundary condition; it vanishes at each wall. A single mode of frequency ω which satisfies Maxwell's equations and the periodic cavity boundary condition can be written, (Sargent, Scully and Lamb 1974), as

$$E_x(z,t) = (2\omega^2 m/V\varepsilon_0)^{\frac{1}{2}} \, q(t) \, \sin kz \tag{2.5}$$

where V is the cavity volume, the wavevector $\underline{k} = (\omega/c)\hat{\underline{z}}$, q(t) is a time-dependent envelope with dimensions of length and m a constant of dimension of mass. This combination of factors may appear odd but can be justified by what follows. The magnetic field in the cavity can be obtained from eq. (2.5) using eq. (2.1) to give

$$H_y(z,t) = (\varepsilon_0/k) \, (2\omega^2 m/V\varepsilon_0)^{\frac{1}{2}} \, \dot{q}(t) \, \cos kz \ . \tag{2.6}$$

We see that the time-dependence of $\underline{E}$ is given by a factor $q(t)$ of dimensions of length whereas that of $\underline{B}$ is given by a factor $\dot{q}(t)$ with dimensions of velocity.

The classical field Hamiltonian or energy of the single-mode field is

$$\mathcal{H} = \frac{1}{2}\int dV(\varepsilon_o \underline{E}_x^2(z,t) + \mu_o H_y^2(z,t)) \tag{2.7}$$

which using eqs.(2.5) and (2.6) gives

$$\mathcal{H} = \frac{1}{2} m\omega^2 q^2 + p^2/2m \tag{2.8}$$

where $p \equiv m\dot{q}$. The Hamiltonian in eq.(2.8) is immediately identifiable as that of a harmonic oscillator of mass m, of frequency ω and position coordinate $q(t)$. As $E_x(z,t) \propto q$ and $H_y(z,t) \propto \dot{q}(= p/m)$, we see that the single field mode behaves as a classical harmonic oscillator with canonical position and momentum variables described by $E_x(z,t)$ and $H_y(z,t)$.

2.2 QUANTIZATION OF A SINGLE-MODE FIELD

We saw in Section (2.1) that a single mode of the radiation field behaves like a simple harmonic oscillator and we shall assume familiarity with the quantization of material oscillators (e.g. Matthews 1974, Merzbacher 1961). The canonical position and momentum variables p and q are replaced by the non-commuting Hermitian operators $\hat{p}$ and $\hat{q}$ which obey the basic commutation relation

$$[\hat{q},\hat{p}] \equiv \hat{q}\hat{p} - \hat{p}\hat{q} = i\hbar \tag{2.9}$$

and the oscillator is described by the Hamiltonian

$$\hat{\mathcal{H}} = \frac{1}{2} m\omega^2\hat{q}^2 + \hat{p}^2/2m \ . \tag{2.10}$$

The electric field operator $\hat{E}_x(z,t)$ is now proportional to the canonical position operator $\hat{q}$

$$\hat{E}_x(z,t) = (2\omega^2 m/V\varepsilon_o)^{\frac{1}{2}}\ \hat{q}(t)\ \sin kz \tag{2.11}$$

and the magnetic field operator to the momentum operator $\hat{p}$

$$\hat{H}_y(z,t) = (\varepsilon_o/k)\ (2\omega^2 m/V\varepsilon_o)^{\frac{1}{2}}\ (\hat{p}/m)\ \cos kz \ . \tag{2.12}$$

It is important to recognise how little is formally altered on quantization. In particular notice how the geometry of the system, which gives rise to the boundary conditions which will be crucial in interference and diffraction, determines the spatial factors which are common in both classical and quantum representation of the field. Quantum optics stems from the non-commutability of the operators. The quantum laws of probability determine the amplitude of the mode excitation.

It is convenient to introduce non-Hermitian combinations of $\hat{p}$ and $\hat{q}$

$$\hat{a} = (2m\hbar\omega)^{-\frac{1}{2}} (m\omega\hat{q} + i\hat{p}) \tag{2.13}$$

$$\hat{a}^+ = (2m\hbar\omega)^{-\frac{1}{2}} (m\omega\hat{q} - i\hat{p}) , \tag{2.14}$$

so that the quantized field Hamiltonian can be written as

$$\hat{\mathcal{H}} = \hbar\omega(\hat{a}^+\hat{a} + \frac{1}{2}) \tag{2.15}$$

and the canonical commutation relation as

$$[\hat{a},\hat{a}^+] = 1 . \tag{2.16}$$

The operator $\hat{a}$ is called the annihilation, or destruction operator. The operator $\hat{a}^+$ is called the creation operator. Using these operators it follows that we can write the electric field operator for the mode of frequency ω as

$$\hat{E}_x(z,t) = \mathcal{E}(\hat{a} + \hat{a}^+) \sin kz \tag{2.17}$$

where $\mathcal{E}$, given by

$$\mathcal{E} = (\hbar\omega/V\varepsilon_0)^{\frac{1}{2}} , \tag{2.18}$$

is an amplitude which will turn out to be the "electric field per photon" (Sargent, Scully and Lamb 1974). The time-dependence of the annihilation operator $\hat{a}$ and the creation operator $\hat{a}^+$ can be determined using Heisenberg's equation of motion

$$\frac{d}{dt}\hat{a} = \frac{i}{\hbar}[\hat{\mathcal{H}}, \hat{a}] . \tag{2.19}$$

The commutator can be evaluated using eqs.(2.15) and (2.16)

$$[\hat{\mathcal{H}}, \hat{a}] = [\hbar\omega(\hat{a}^+\hat{a} + \frac{1}{2}), \hat{a}] = \hbar\omega(\hat{a}^+\hat{a}\hat{a} - \hat{a}\hat{a}^+\hat{a})$$

$$= \hbar\omega(\hat{a}^+\hat{a} - \hat{a}\hat{a}^+)\hat{a} = -\hbar\omega\hat{a} \tag{2.20}$$

So

$$\frac{d}{dt}\hat{a} = - i\omega\hat{a} \tag{2.21}$$

which has the solution

$$\hat{a}(t) = \hat{a}(0)e^{-i\omega t}. \tag{2.22}$$

By the same method we find

$$\hat{a}^+(t) = \hat{a}^+(0)e^{i\omega t} . \tag{2.23}$$

The product $\hat{a}^+\hat{a}$ has a special significance and is called the number operator $\hat{n}$. Let $|n\rangle$ be an energy eigenstate of the single-mode field with energy eigenvalue E_n, such that

$$\hat{\mathcal{H}}|n\rangle = \hbar\omega(\hat{a}^+\hat{a} + \frac{1}{2})|n\rangle = E_n|n\rangle \ . \tag{2.24}$$

We can then generate a new eigenvalue equation by multiplying eq.(2.24) from the left by $\hat{a}^+$ to give

$$\hbar\omega(\hat{a}^+\hat{a}^+\hat{a} + \frac{1}{2}\hat{a}^+)|n\rangle = E_n\hat{a}^+|n\rangle \ . \tag{2.25}$$

Using the commutator (eq.(2.16)) this can be re-written as

$$\hbar\omega[(\hat{a}^+\hat{a} - 1)\hat{a}^+ + \frac{1}{2}\hat{a}^+)|n\rangle = E_n\hat{a}^+|n\rangle$$

or

$$\hbar\omega(\hat{a}^+\hat{a} + \frac{1}{2})(\hat{a}^+|n\rangle) = (E_n + \hbar\omega)(\hat{a}^+|n\rangle) \ , \tag{2.26}$$

which is an eigenvalue equation for the eigenstate $(\hat{a}^+|n\rangle)$ with eigenenergy $(E_n + \hbar\omega)$. The application of the creation operator $\hat{a}^+$ has resulted in a new state containing one more quantum of excitation. We may denote this new state as $|n+1\rangle$, but it needs to be normalised. In a similar fashion we find that the operator $\hat{a}$ destroys one unit of excitation, generating a new eigenstate $|n-1\rangle$ with energy $(E_n - \hbar\omega)$. This cannot be repeated indefinitely since the system has a ground state $|0\rangle$ of energy E_0 which must be positive. So we demand

$$\hat{\mathcal{H}}(\hat{a}|n\rangle) = (E_n - \hbar\omega)(\hat{a}|n\rangle)$$

$$\hat{\mathcal{H}}(\hat{a}|0\rangle) = (E_0 - \hbar\omega)(\hat{a}|0\rangle) \ , \tag{2.27}$$

which defines the ground state through

$$\hat{a}|0\rangle = 0 \ . \tag{2.28}$$

Therefore the eigenvalue equation for the ground state is

$$\hat{\mathcal{H}}|0\rangle = \hbar\omega(\hat{a}^+\hat{a} + \frac{1}{2})|0\rangle = \frac{1}{2}\hbar\omega|0\rangle \tag{2.29}$$

so that the ground state of the single mode electromagnetic field has a non-zero energy $\hbar\omega/2$. This zero-point energy reflects our inability to confine a "particle", which here is the mode "oscillator", at rest and still satisfy the Heisenberg uncertainty relation. Since $E_{n+1} = E_n + \hbar\omega$,

$$E_n = \hbar\omega(n + \frac{1}{2}), \ n = 0,1,2\ldots \tag{2.30}$$

and

$$\hat{n}|n\rangle = \hat{a}^+\hat{a}|n\rangle = n|n\rangle \ . \tag{2.31}$$

The "number states" $|n\rangle$ are normalised, with

$$\langle n-1|n-1\rangle = 1 = \langle n|n\rangle = \langle n+1|n+1\rangle \ . \tag{2.32}$$

If we write

$$\hat{a}|n\rangle = C_n|n-1\rangle \tag{2.33}$$

then

$$(\langle n|\hat{a}^+)(\hat{a}|n\rangle) = \langle n-1|C_n^*C_n|n-1\rangle = n$$

so that

$$|C_n|^2 = n\ .$$

The constant C_n can be taken as real without loss of generality, so that

$$C_n = n^{\frac{1}{2}} \tag{2.34}$$

and

$$\hat{a}|n\rangle = \sqrt{n}\ |n-1\rangle \tag{2.35}$$

$$\hat{a}^+|n\rangle = \sqrt{n+1}\ |n+1\rangle\ . \tag{2.36}$$

The spectrum of the radiation field is thus of a ladder of equally spaced levels separated by $\hbar\omega$, which one ascends by the action of $\hat{a}^+$ and descends by the action of $\hat{a}$. The number states are generated by the repeated action of the creation operator $\hat{a}^+$:

$$|n\rangle = (\sqrt{n!})^{-1}\ (\hat{a}^+)^n|0\rangle\ . \tag{2.37}$$

The only non-vanishing matrix elements of the annihilation and creation operators are

$$\langle n-1|\hat{a}|n\rangle = \langle n-1|\sqrt{n}\ |n-1\rangle = \sqrt{n} \tag{2.38}$$

$$\langle n+1|\hat{a}^+|n\rangle = \langle n+1|\sqrt{n+1}\ |n+1\rangle = \sqrt{n+1}\ . \tag{2.39}$$

2.3 QUANTUM FLUCTUATIONS OF A SINGLE-MODE FIELD

The number states $|n\rangle$ are energy eigenstates of the single mode field. But they do not describe a field with a well-defined electromagnetic field, as

$$\langle n|\hat{E}_x(z,t)|n\rangle = \mathcal{E}\sin kz(\langle n|\hat{a}(t)|n\rangle + \langle n|\hat{a}^+(t)|n\rangle)$$
$$= 0 \tag{2.40}$$

using eqs.(2.38) and (2.39). This does not, of course, mean that the field is zero as,

$$\langle n|\hat{E}_x^2(z,t)|n\rangle = \mathcal{E}^2\sin^2 kz\ \langle n|(\hat{a}^+\hat{a}^+ + \hat{a}\hat{a}^+ + \hat{a}^+\hat{a} + \hat{a}\hat{a})|n\rangle$$
$$= 2\mathcal{E}^2\sin^2 kz\ (n+\tfrac{1}{2})\ . \tag{2.41}$$

The r.m.s. electric field in the cavity when only one photon is present ignoring the ever present zero-point-energy is $\sqrt{2}\mathcal{E}\sin kz$. When this is averaged over spatial position z we see that the "electric field per photon" is indeed $\mathcal{E}$. The number states field can be visualized as having a definite amplitude with a phase randomly distributed over 2π (Loudon 1973). The field energy, or number of quanta n, and the field strength are complementary concepts (Power 1964), because the number operator $\hat{n}$ does not commute with the electric field operator $\hat{E}_x(z,t)$. We see that

$$[\hat{n},\hat{E}_x] = \mathcal{E}\, \sin kz(\hat{a}^+ - \hat{a}) = -\, 2i\hat{p}(2mV\varepsilon_0)^{\frac{1}{2}}\sin kz\ . \tag{2.42}$$

If the excitation number n is known exactly the fluctuations in $E_x(z,t)$ are of the order of $E_x(z,t)$. Conversely if $E_x(z,t)$ is known accurately the excitation number n is correspondingly uncertain.

2.4 MULTIMODE FIELDS

So far we have studied only a single mode of a field confined to a cavity. These results can be generalized to describe a multi-mode radiation field. Before we do this, we stress that the quanta in the cavity of energy $\hbar\omega$ are the photons of the system, and as such are not localized particles but are characteristic excitations of the cavity mode and spread over the mode volume. Quantum mechanics introduces not localized corpuscular photons as in the old quantum theory, but rather mode-excitations of the system.

In free space, the electric and magnetic radiation fields can be described in terms of the vector potential $\underline{A}$ which obeys the wave equation

$$\nabla^2\underline{A} - \frac{1}{c^2}\frac{\partial^2\underline{A}}{\partial t^2} = 0 \tag{2.43}$$

where

$$\underline{B} = \underline{\nabla} \times \underline{A}$$

and

$$\underline{E} = -\frac{\partial}{\partial t}\underline{A}\ .$$

We take running wave solutions to the wave equation of the plane-wave form

$$\underline{A} = \sum_{\underline{k}} \left\{ \underline{A}_{\underline{k}}(t)e^{i\underline{k}.\underline{r}} + \underline{A}^*_{\underline{k}}(t)e^{-i\underline{k}.\underline{r}} \right\} \tag{2.44}$$

subject to the periodic boundary conditions (Loudon 1973)

$$k_x = \frac{2\pi}{L}m_x\ ,\quad k_y = \frac{2\pi}{L}m_y\ ,\quad k_z = \frac{2\pi}{L}m_z \tag{2.45}$$

where the "box" volume $V = L^3$ and m_x, m_y, $m_z = 0, \pm1, \pm2$---. The mode density, the number of modes of wavenumber between k and k + dk, is

$$\rho(k)dk = k^2dk/\pi^2\ . \tag{2.46}$$

Alternatively in frequency space the number of modes of frequency between ω and $\omega + d\omega$ is

$$\rho(\omega)d\omega = \omega^2 d\omega/\pi^2 c^3 \ . \tag{2.47}$$

In free space, the discrete sum in eq.(2.44) is replaced by a continuum integral

$$\sum_k \longrightarrow \frac{V}{\pi^2}\int k^2 dk \ . \tag{2.48}$$

In quantum optics the Coulomb gauge, in which $\underline{\nabla}.\underline{A} = 0$, is mainly used. In this gauge, $\underline{k}.\underline{A}_k(t) = 0 = \underline{k}.\underline{A}_k^*(t)$ and each Fourier component satisfies, from eq.(2.43), the harmonic oscillator equation

$$\frac{\partial^2 \underline{A}_k}{\partial t^2} + \omega_k^2 \underline{A}_k = 0 \tag{2.49}$$

where $\omega_k = ck$. From eq.(2.49) we find

$$\underline{A}_k(t) = \underline{A}_k \, e^{-i\omega_k t} . \tag{2.50}$$

The cycle-averaged energy of mode $\underline{k}$ is

$$\mathcal{E}_k = \frac{1}{2}\int (\varepsilon_o \overline{\underline{E}_k^2} + \mu_o \overline{\underline{H}_k^2}) dV \tag{2.51}$$

where the bar denotes the cycle average. The electric field from eq.(2.43) is,

$$\underline{E}_k = i\omega_k \left\{ \underline{A}_k \, e^{-i\omega_k t + i\underline{k}.\underline{r}} - \underline{A}_k^* \, e^{i\omega_k t - i\underline{k}.\underline{r}} \right\} \tag{2.52}$$

and the cavity magnetic field is

$$\underline{H}_k = (i/\mu_o)\underline{k} \times \left\{ \underline{A}_k \, e^{-i\omega_k t + i\underline{k}.\underline{r}} - \underline{A}_k^* \, e^{i\omega_k t - i\underline{k}.\underline{r}} \right\} \tag{2.53}$$

so that

$$\overline{\mathcal{E}}_k = 2\varepsilon_o V \omega_k^2 \, \underline{A}_k . \underline{A}_k^* \ . \tag{2.54}$$

The vector potential can be written in terms of generalized canonical position q_k and momentum p_k variables

$$\underline{A}_k = (4\varepsilon_o V \omega_k^2)^{-\frac{1}{2}} (\omega_k q_k + i p_k) \underline{\varepsilon}_k \tag{2.55}$$

$$\underline{A}_k^* = (4\varepsilon_o V \omega_k^2)^{-\frac{1}{2}} (\omega_k q_k - i p_k) \underline{\varepsilon}_k \tag{2.56}$$

where $\underline{\varepsilon}_k$ is the polarization unit vector. In terms of these variables the cycle-averaged energy is written as

$$\overline{\mathcal{E}}_k = \frac{1}{2}(p_k^2 + \omega_k^2 q_k^2) , \tag{2.57}$$

the energy of a unit mass simple harmonic oscillator. The complete free field Hamiltonian is given by the sum of all single-mode terms given by eq. (2.57). The summation is over the wavevectors $\underline{k}$ and polarization directions $\underline{\varepsilon}_k$.

Each mode $\underline{k}$, represented as a unit mass harmonic oscillator, can be quantized in terms of number states as before. The state of the multimode field can be written as a product of the individual mode states,

$$|\{n_k\}\rangle = |n_{k1}\rangle|n_{k2}\rangle---|n_{kj}\rangle---$$

$$= |n_{k1}, n_{k2}, --- n_{kj}, ---\rangle , \tag{2.58}$$

where n_{kj} represents the number of quanta in mode with wavevector $\underline{k}_j$. The creation operator $\hat{a}_{k1}$ acts on the multimode state

$$\hat{a}^+_{k1} | n_{k1}, n_{k2}, ---n_{kj}, ---\rangle = (n_{k1} + 1)^{\frac{1}{2}} | n_{k1} + 1, n_{k2}, -- n_{kj}, ---\rangle. \tag{2.59}$$

The field amplitudes $\underline{A}_k$ of classical theory become

$$\hat{\underline{A}}_k = (\hbar/2\varepsilon_o V\omega_k)^{\frac{1}{2}} \hat{a}_k \hat{\underline{\varepsilon}}_k \tag{2.60}$$

and

$$\hat{\underline{A}}^*_k = (\hbar/2\varepsilon_o V\omega_k)^{\frac{1}{2}} \hat{a}^+_k \hat{\underline{\varepsilon}}_k \tag{2.61}$$

such that the vector field operator can be written as

$$\hat{\underline{A}} = \sum_{\underline{k}} (\hbar/2\varepsilon_o V\omega_k)^{\frac{1}{2}} \underline{\varepsilon}_k \left\{ \hat{a}_k e^{-i\omega_k t + i\underline{k}.\underline{r}} + \hat{a}^+_k e^{i\omega t - i\underline{k}.\underline{r}} \right\} . \tag{2.62}$$

(2.62)

The electric field operator is

$$\hat{\underline{E}}(\underline{r},t) = \sum_{\underline{k}} i(\hbar\omega_k/2\varepsilon_o V)^{\frac{1}{2}} \underline{\varepsilon}_k \left\{ \hat{a}_k(t) e^{-i\omega_k t + i\underline{k}.\underline{r}} - \hat{a}^+_k(t) e^{i\omega_k t - i\underline{k}.\underline{r}} \right\} \tag{2.63}$$

and the magnetic field operator

$$\hat{\underline{H}}(\underline{r},t) = \sum_{\underline{k}} i(\hbar c^2/2\mu_o V\omega_k)^{\frac{1}{2}} \underline{k} \times \underline{\varepsilon}_k \left\{ \hat{a}_k(t) e^{-i\omega_k t + i\underline{k}.\underline{r}} - \hat{a}^+_k(t) e^{i\omega_k t - i\underline{k}.\underline{r}} \right\}. \tag{2.64}$$

The quantum expression for a single-mode plane wave field from eq.(2.63) is

$$\hat{\underline{E}}(r,t) = i(\hbar\omega/2\varepsilon_0 V)^{\frac{1}{2}} \underline{\varepsilon}_k \left\{ \hat{a}_k e^{-i\omega_k t + i\underline{k}.\underline{r}} - \hat{a}_k^+ e^{i\omega_k t - i\underline{k}.\underline{r}} \right\}. \quad (2.65)$$

In much of quantum optics the spatial variation of the electromagnetic field mode over the dimensions of the atomic system may be neglected. In this case the exponential factors $e^{\pm i\underline{k}.\underline{r}}$ may be set equal to unity. This single-mode expression can be used either in the interaction representation in which the operator $\hat{a}(t)$ and its conjugate $\hat{a}^+(t)$ have the time development shown in eq. (2.65) or in the Schrödinger representation in which $\hat{a}$ and $\hat{a}^+$ have no time development. The expressions for the electric field operator in these two pictures coincide at $t = 0$.

2.5 ZERO-POINT ENERGY AND VACUUM FLUCTUATIONS

In Section (2.3) we saw that the quantized radiation field fluctuates. For a single-mode field described by a number state $|n\rangle$, the r.m.s. deviation in electric field strength $\Delta E = (\langle E^2\rangle - \langle E\rangle^2)^{\frac{1}{2}}$, using eq.(2.40) and (2.41) is

$$\Delta E = \sqrt{2}\mathcal{E}\sqrt{n + \frac{1}{2}}\ \sin kz\ . \quad (2.66)$$

If the single-mode field is unoccupied, $n = 0$, then the r.m.s. fluctuation in the vacuum field strength is $\Delta E(\text{vac}) = \mathcal{E}\sin kz$. Vacuum fluctuations and zero-point energy have a common origin in the non-commutability of the field annihilation and creation operators $\hat{a}$ and $\hat{a}^+$. The occurrence of the zero-point energy term and its associated vacuum fluctuation presents severe problems in quantum field theory. In practice there are an infinite number of radiation field modes, each with a finite zero-point energy. The total zero-point energy therefore diverges unless the high frequency modes are excluded from the sum. Yet the zero-point energy seems to lead to observable consequences (Casimir 1948, Power 1964) and cannot be ignored. A stimulating analysis of the role of vacuum fluctuations may be found in the article by Jaynes (1978) in the proceedings of the 1977 Rochester Coherence and Quantum Optics conference.

The major observable effect attributed to the existence of vacuum fluctuations is the Lamb shift. In 1947, Lamb and Retherford (Reprinted Paper 4) used a microwave frequency method to examine the fine structure of the $n = 2$ level of atomic hydrogen. Earlier high resolution optical studies of the H_α line seemed to indicate a discrepancy between experiment and the Dirac relativistic theory of the hydrogen atom. The Dirac theory predicts that the $2^2S_{1/2}$ and $2^2P_{1/2}$ levels should be degenerate. The early optical work suggested that these states were not in fact degenerate but separated by about 0.033 cm^{-1}. Lamb and Retherford used an elegant combination of atomic beam and microwave techniques and showed that the $2^2S_{1/2}$ state is higher in energy than the $2^2P_{1/2}$ state by about 1000 MHz. The lifting of the $2^2P_{1/2}$ and $2^2S_{1/2}$ degeneracies is explained by the interaction of the bound electron with the vacuum fluctuations (Bethe 1947). In the paper by Power (Reprinted Paper 5), the Lamb shift, or radiative level shift of the $2^2S_{1/2}$ level, is calculated directly from the change in zero-point energy due to the interaction of the vacuum field with the hydrogen atom.

A simple intuitive interpretation of the Lamb shift was given by Welton (1948). Each field mode contains $h\nu/2$ zero-point energy. The number of modes in a cavity of volume V with frequency between ν and $\nu + d\nu$ is $(8\pi/c^3)\nu^2 d\nu V$. The zero-point field energy is

$$\left(\frac{8\pi}{c^3}\nu^2 d\nu V\right)\frac{1}{2}h\nu = \frac{1}{8\pi}\int_V (E_\nu^2 + B_\nu^2)\,dV = \frac{1}{8\pi}E_\nu^2 V \tag{2.67}$$

where E_ν is the amplitude of the electric field component of frequency ν. The square of the vacuum electric field is therefore

$$E_\nu^2 = \frac{32\pi^2}{c^3}h\nu^3 d\nu\ . \tag{2.68}$$

The electron bound in the hydrogen atom interacts with the fluctuating zero-point electric field and with the Coulomb potential of the proton $-e^2/r$. The perturbation of the electron from its standard "orbit" is described by the fluctuating electron position Δr

$$r \rightarrow r + \Delta r\ . \tag{2.69}$$

The change in potential energy is $\Delta V = V(r + \Delta r) - V(r)$, which by Taylor's theorem is

$$\Delta V = \Delta x\frac{\partial V}{\partial x} + \Delta y\frac{\partial V}{\partial y} + \Delta z\frac{\partial V}{\partial z} + \frac{1}{2}(\Delta x)^2\frac{\partial^2 V}{\partial x^2} + \frac{1}{2}(\Delta y)^2\frac{\partial^2 V}{\partial y^2} + \frac{1}{2}(\Delta z)^2\frac{\partial^2 V}{\partial z^2} + \cdots .$$

The fluctuations are isotropic, so that $\langle\Delta x\rangle = \langle\Delta y\rangle = \langle\Delta z\rangle = 0$ and $\langle(\Delta x)^2\rangle = \langle(\Delta y)^2\rangle = \langle(\Delta z)^2\rangle = \langle(\Delta r)^2\rangle/3$. Then

$$\langle\Delta V\rangle = \frac{1}{6}\langle(\Delta r)^2\rangle\nabla^2 V\ . \tag{2.70}$$

The perturbation $\langle\Delta V\rangle$ changes the energy of an atomic state $|n\ell m_\ell\rangle$ by an amount Δ. To first order the energy shift is

$$\Delta = \langle n\ell m_\ell|(\langle\Delta V\rangle)|n\ell m_\ell\rangle \tag{2.71}$$

$$= \frac{1}{6}\langle(\Delta r)^2\rangle\langle n\ell m_\ell|\nabla^2 V|n\ell m_\ell\rangle\ . \tag{2.72}$$

Using $V(r) = -e^2/r$ and $\nabla^2(1/r) = -4\pi\delta(r)$, we find

$$\langle n\ell m_\ell|\nabla^2 V|n\ell m_\ell\rangle = 4\pi e^2\,|\psi_{n\ell m_\ell}(r=0)|^2\ . \tag{2.73}$$

Non-relativistic atomic wavefunctions vanish at the origin except for s-states with $\ell = 0$, where

$$|\psi_{noo}(r=0)|^2 = 1/\pi n^3 a_o^3 \tag{2.74}$$

and a_o is the Bohr radius. For p-states the wavefunction vanishes and hence so does the energy shift. The mean square displacement $\langle(\Delta r)^2\rangle$ is obtained by assuming that the important field frequencies greatly exceed the atomic resonance frequencies; lower frequencies are shielded by the atomic binding and cannot influence the motion of the electron. The electron-field interaction leads to an equation of motion for the displacement induced by the fluctuating field. The displacement induced with frequency between ν and $\nu + d\nu$ is determined by

$$\frac{d^2}{dt^2}(\Delta r_\nu) = \frac{eE_\nu}{m} \exp(2\pi i\nu t) \tag{2.75}$$

with the solution

$$\Delta r_\nu = -\frac{e}{m}\frac{E_\nu}{4\pi^2\nu^2} \exp(2\pi i\nu t) \ . \tag{2.76}$$

The mean square displacement induced by these modes is

$$<(\Delta r_\nu)^2> = \frac{e^2}{m^2}\ \frac{E_\nu^2}{32\pi^4\nu^4} = \frac{e^2 h}{\pi^2 m^2 c^3}\ \frac{d\nu}{\nu} \tag{2.77}$$

using eq.(2.68). We obtain the energy shift for s-states from eqs.(2.72) - (2.77) summed over all frequencies.

$$\Delta = \frac{2}{3}\left(\frac{e^2}{\hbar c}\right)^2 \left(\frac{\hbar}{mc}\right)^2 \ \frac{\hbar c}{\pi^2 n^3 a_o^3} \int \frac{d\nu}{\nu} \tag{2.78}$$

where $(e^2/\hbar c) = \alpha$ the fine structure constant, and $(\hbar/mc) = \lambda_c$ the Compton wavelength of the electron. The divergent frequency integral is cut off at both high and at low frequencies. At low frequencies the atom does not respond to the fluctuating electric field and a natural cut-off is the frequency of the electron in its orbit, $\nu_0 = e^2/\hbar a_o^3 n^3$. The analysis also breaks down at high frequencies where relativistic effects affect the electron's motion. The preceding analysis is limited to velocities $v \ll c$ (Power 1964),

$$\frac{v}{c} = \frac{(p/m)}{c} = \frac{pc}{mc^2} = \frac{\hbar k}{mc} \lesssim 1$$

which restrict wavenumber k to less than $(mc/\hbar)$ and angular frequencies to less than $(mc^2/\hbar)$ in the integral in eq.(2.78). For the $2^2S_{1/2}$ state in hydrogen using $a_0 = \hbar^2/me^2$, the energy shift is

$$\Delta = \frac{1}{6\pi}\alpha^3 \frac{me^4}{\hbar^2} \log\left(\frac{mc^2}{h\nu_o}\right) , \tag{2.79}$$

giving $\Delta \sim 1000$ MHz. The $2^2P_{1/2}$ state is unaffected to this order by the radiative corrections leaving the Lamb shift $\Delta(2^2S_{1/2}) - \Delta(2^2P_{1/2}) \approx 1000$ MHz. A review of the current status of the Lamb shift theory and experiment is to be found in the paper by Newton et al. (1979). As well as shifting the atomic levels, the fluctuating vacuum field can be thought of an "inducing" spontaneous decay from excited atomic states (Schiff 1955, Milonni 1976).

The fully-quantized approach outlined in this chapter can be used to derive the wave and particle parts of the Einstein fluctuation formula in a very simple way. The mean square photon number fluctuation of a single-mode field is

$$<(\Delta n)^2> = <\hat{a}^+\hat{a}\hat{a}^+\hat{a}> - <\hat{a}^+\hat{a}>^2 \ . \qquad (2.80)$$

The first term on the right hand side of eq.(2.80) can be written in normal order, that is with creation operators to the left, annihilation operators to the right, using the commutator $[\hat{a},\hat{a}^+] = 1$. We find

$$\begin{aligned} <(\Delta n)^2> &= <\hat{a}^+(\hat{a}^+\hat{a} + 1)\hat{a}> - <\hat{a}^+\hat{a}>^2 \\ &= <\hat{a}^+\hat{a}> + (<\hat{a}^+\hat{a}^+\hat{a}\hat{a}> - <\hat{a}^+\hat{a}>^2) \\ &= <n> + <(\Delta n)^2>_{\text{waves}} \ . \end{aligned} \qquad (2.81)$$

The second term has the character of the fluctuations in intensity of a classical wave. This is the only term present in a classical analysis where $\hat{a}$ and $\hat{a}^+$ are taken to commute and the classical wave amplitude is proportional to $\hat{a}$. The first or particle-like term in eq.(2.81) arises from the noncommutability of $\hat{a}$ and $\hat{a}^+$ (or $\hat{p}$ and $\hat{q}$). The single-mode zero-point energy is also attributed to this noncommutability

$$\begin{aligned} \hat{\mathcal{H}} &= \frac{1}{2} (\hat{p}^2 + \omega^2\hat{q}^2) = \frac{1}{2}\hbar\omega \ (\hat{a}^+\hat{a} + \hat{a}\hat{a}^+) \\ &= \hbar\omega \ (\hat{a}^+\hat{a} + \frac{1}{2}) \end{aligned} \qquad (2.82)$$

and so, as noticed first by Born, Heisenberg and Jordan (1926), the zero-point energy and the particle fluctuation term in $<(\Delta n)^2>$ are closely related.

2.6 MODE OCCUPATION AND PHOTONS

The electric field operator in an arbitrary enclosure can be written, generalising eq.(2.63) as

$$\underline{\hat{E}}(\underline{r},t) = i \sum_{\underline{k}} (\hbar\omega_k)^{\frac{1}{2}} [\hat{a}_k(t)\underline{F}_k(\underline{r}) - \hat{a}_k^+(t)\underline{F}_k^*(\underline{r})] \qquad (2.83)$$

where the mode function $\underline{F}_k(\underline{r})$ satisfies the wave-equation subject to the usual boundary conditions used in the classical theory. Interference effects are determined by the spatial mode functions $\underline{F}_k(\underline{r})$ and are precisely those of classical theory. The size of the interference effects however depends on the expectation value of combinations of field mode annihilation and creation operators. For example, the intensity distribution depends on the probability of detecting a *single* photon at position $\underline{r}$ and time t. This involves terms of the form $<\hat{a}_k^+(t)\hat{a}_k(t)>|\underline{F}_k(r)|^2$ and cross-terms between the different contributing modes. The interference pattern is built up from a superposition of one-photon probability patterns (Taylor (Reprinted Paper 1)). Pfleegor and Mandel (Reprinted Paper 6) and Magyar and Mandel (Reprinted Paper 7) have demonstrated that the fields from the independent lasers produce interference fringes. This may be difficult to understand using a crude billiard ball interpretation of the "photon". Since however the photon is an occupation of one of the normal modes of the whole system including both lasers, the quantum explanation is straightforward. The detector measures the occupation of normal modes. In this experiment there is no way of telling from which laser the energy derived and fringes would be expected.

If it is known that only a single photon is present in the apparatus at any one time, an attempt to divide the photon with a beam splitter can be made. Photon division is precluded by quantum theory but is allowed by semi-classical radiation theory, or by any theory in which the photon is viewed as a packet of electromagnetic energy. This experiment was performed by Clauser in 1974 (Reprinted Paper 8) who found no evidence that photons can be split, in clear contradiction with semi-classical theory. Such disagreements between semi-classical and quantum theories of coherence are rare: in Chapter 6 we discuss them further.

CHAPTER 3

Absorption and Emission of Radiation

3.1 INTERACTION OF AN ATOM WITH A RADIATION FIELD

In this Chapter we review the application of quantum mechanical perturbation theory to the absorption and emission of electromagnetic radiation by atoms. For details, the reader should refer to a standard quantum mechanics text (e.g. Merzbacher 1961, Matthews 1974). We are concerned here with drawing out certain important features which play a key role in quantum optics. Suppose at time $t < 0$ that we have an unperturbed atomic system described by the Schrödinger equation

$$H_o|k\rangle = \hbar\omega_k|k\rangle. \tag{3.1}$$

We will assume that the eigenstates $|k\rangle$ and their energies $\hbar\omega_k$ are in some way known, and that the atom is prepared in a particular state at $t \le 0$. For most situations this will be the atomic ground state which we will label $|i\rangle$. Then at time t_o we imagine the interaction between the atom and a classical electromagnetic field is switched on. For example a light pulse might be injected into a vapour cell containing the atoms, or an atomic beam intersects a beam of optical radiation. In this way the interaction can be switched on and off in a controlled way. We will calculate in this section the probability of the incident radiation exciting a transition to an excited state of the atom.

The coupling of an atom to an electromagnetic field can be described in a number of different ways. The radiation field can be described in terms of the vector potential $\underline{A}$ which couples to the atomic electron through the "minimal substitution" Hamiltonian (Power 1964)

$$\hat{\mathcal{H}} = \frac{1}{2m}\,(\hat{\underline{p}} - e\underline{A}(\underline{r},t))^2 + \hat{V}(r) \tag{3.2}$$

where $V(r)$ is the Coulomb interaction which binds the electron to the atomic nucleus. The unperturbed part of the Hamiltonian is

$$\hat{\mathcal{H}}_o = \frac{\hat{\underline{p}}^2}{2m} + \hat{V}(r) \tag{3.3}$$

and the external perturbation due to the radiative interaction is

$$\hat{\mathcal{H}}' = \hat{\mathcal{H}}_1 + \hat{\mathcal{H}}_2 = -\frac{e}{m}\,\hat{\underline{p}}.\underline{A}(\underline{r}) + \frac{e^2}{2m}\,\underline{A}^2(\underline{r}), \tag{3.4}$$

where the external radiation is described in the Coulomb gauge by the wave-equation

$$\nabla^2 \underline{A} - \frac{1}{c^2}\frac{\partial^2 \underline{A}}{\partial t^2} = 0 \tag{3.5}$$

with zero static potential

$$\phi = 0 \ , \quad \nabla.\underline{A} = 0 \ . \tag{3.6}$$

The first term in eq.(3.4), $\hat{\mathcal{H}}_1 = -(e/m)\hat{\underline{p}}.\underline{A}(\underline{r})$ will be responsible for exciting the atom. The second term is quadratic in the vector potential and because it is independent of the atomic state will not excite transitions and will be ignored. The minimal substitution form of the interaction is widely used in quantum optics (see for example the paper by Mandel, Sudarshan and Wolf, Reprinted Paper 9).

An alternative form of the interaction Hamiltonian describing the coupling of the atom to the radiation field is the electric dipole interaction

$$\hat{V} = -\ \hat{\underline{d}}.\underline{E} \tag{3.7}$$

where $\hat{\underline{d}}$ is the electric dipole operator $\hat{\underline{d}} = e\hat{\underline{r}}$ and $\underline{E} = -\dot{\underline{A}}$ is the electric field of the radiation. The two forms of the interaction are equivalent for our purposes (see e.g. Power and Zienau 1959; Power 1964; Sargent, Scully and Lamb 1974). We will use the electric-dipole form eq.(3.7) from now on.

The presence of the interaction term $\hat{V}$ which is turned on at $t = 0$ mixes together the eigenstates $|k\rangle$ of $\hat{\mathcal{H}}_0$. Although we prepared the system in state $|i\rangle$ at $t = 0$, at later times we will find that transitions to other states have occurred. The wavefunction $\psi(t)$ of the driven atomic system may be expanded in terms of the complete set of uncoupled atomic states $|k\rangle$

$$|\psi(t)\rangle = \sum_k a_k(t) e^{-i\omega_k t} |k\rangle \tag{3.8}$$

where $\omega_k = E_k/\hbar$. The time-dependent expansion coefficients $a_k(t)$ describe how the coupling V mixes the unperturbed states. They are determined from the Schrödinger equation for the time-development of $|\psi(t)\rangle$ and the use of the property of orthonormality. The wave function of the coupled system is normalised

$$\langle\psi|\psi\rangle = 1. \tag{3.9}$$

Using eq.(3.8) in eq.(3.9) we find using $\langle k|k'\rangle = \delta_{kk'}$

$$\sum_k |a_k(t)|^2 = 1$$

so the $a_k(t)$ factors are just time-dependent probability amplitudes.

The exact equations of motion for these probability amplitudes is obtained by using the wavefunction expansion for $|\psi(t)\rangle$ in the full time-dependent Schrödinger equation

$$\frac{i\hbar\partial|\psi(t)\rangle}{\partial t} = (\hat{\mathcal{H}}_0 + \hat{V})|\psi(t)\rangle. \tag{3.10}$$

Multiplying from the left by $\langle \ell | e^{i\omega_\ell t}$ picks out the amplitude for state $|\ell\rangle$:

$$\dot{a}_\ell(t) = -\frac{i}{\hbar} \sum_k a_k(t) \langle \ell | \hat{V}(t) | k \rangle e^{i\omega_{\ell k} t}. \tag{3.11}$$

The transition frequencies are $\omega_{\ell k} = (E_\ell - E_k)/\hbar$, and for brevity we will write the matrix element $\langle \ell | \hat{V}(t) | k \rangle = V_{\ell k}(t)$. We see from eq.(3.11) that $\hat{V}(t)$ excites transitions between all accessible states $|k\rangle$ to the state $|\ell\rangle$. The initial condition is that the atom is in state $|i\rangle$ at $t = 0$

$$a_i(0) = 1. \tag{3.12}$$

Population is lost from $|i\rangle$ as time goes on so that $a_i(t)$ is diminished by couplings $V_{if}(t)$ to excited states $f \neq i$. Population is increased in the initially unoccupied excited states $|f\rangle$, so that the probability amplitudes for such excited states, $a_f(t)$ increase. The transition probability for a transition from the initial state $|i\rangle$ to a final state $|f\rangle$ is

$$P_{i\to f}(t) = a_f^*(t) a_f(t). \tag{3.13}$$

The evolution equations (3.11) for the probability amplitudes are exactly soluble only for very simple cases. In general we must make approximations concerning the strength of the coupling $\hat{V}(t)$. In perturbation theory we scale $\hat{V}(t)$ as $\lambda\hat{V}(t)$ where λ is a number in the range $0 \leq \lambda \leq 1$. The probability amplitude for being in some state $|\ell\rangle$ is expanded in a power series

$$a_\ell(t) = a_\ell^{(0)}(t) + \lambda a_\ell^{(1)}(t) + \lambda^2 a_\ell^{(2)}(t) + \cdots \tag{3.14}$$

where the superscript denotes the order of the perturbation. If the expansion (3.14) is inserted into the evolution equations (3.11) and the coefficients of λ equated, we find

$$\dot{a}_\ell^{(0)} = 0 \tag{3.15a}$$

$$\dot{a}_\ell^{(1)} = -\frac{i}{\hbar} \sum_k a_k^{(0)} V_{\ell k}(t) e^{i\omega_{\ell k} t} \tag{3.15b}$$

$$\dot{a}_\ell^{(2)} = -\frac{i}{\hbar} \sum_k a_k^{(1)} V_{\ell k}(t) e^{i\omega_{\ell k} t}. \tag{3.15c}$$

We observe that in general, the lower order amplitudes $a_k^{(n-1)}(t)$ drive the higher order amplitudes $a_\ell^{(n)}(t)$,

$$\dot{a}_\ell^{(n)} = -\frac{i}{\hbar} \sum_k a_k^{(n-1)}(t) V_{\ell k}(t) e^{i\omega_{\ell k} t}. \tag{3.16}$$

The central assumption behind the perturbation expansion is that the effect of $\hat{V}(t)$ is small and that the probability amplitudes do not change much from their initial values at $t = 0$, namely

$$a_i(0) = 1, \quad a_f(0) = 0 \ (f \neq i) \tag{3.17a}$$

and

$$a_i(t) \sim 1, \quad a_f(t) \ll 1, (f \neq i). \tag{3.17b}$$

The amplitude for being excited from the initial state $|i\rangle$ to an excited state $|f\rangle$ in first order can be obtained by integrating eq.(3.16) to give the excited state amplitude driven by the lowest order initial state amplitude

$$a_f^{(1)}(t) = -\frac{i}{\hbar}\int_0^t dt' \, V_{fi}(t')e^{i\omega_{fi}t'} a_i^{(0)}(t'). \tag{3.18}$$

The perturbation $\hat{V}(t)$ can excite the atom from $|i\rangle$ to an excited state $|f\rangle$ and by a further interaction from $|f\rangle$ to another excited state $|f'\rangle$. This is described by the second-order term in eq.(3.16) as being driven by all state amplitudes $a_f^{(1)}(t)$ excited in first order from the ground state,

$$a_{f'}^{(2)}(t) = -\frac{i}{\hbar}\sum_f \int_0^t dt' \, V_{f'f}(t')e^{i\omega_{f'f}t'} a_f^{(1)}(t') \tag{3.19}$$

or using eq.(3.18) as

$$a_{f'}^{(2)}(t) = \sum_f \left(-\frac{i}{\hbar}\right)^2 \int_0^t dt' \int_0^{t'} dt'' V_{f'f}(t')e^{i\omega_{f'f}t'} V_{fi}(t'')e^{i\omega_{fi}t''} a_i^{(0)}(t''). \tag{3.20}$$

This form directly exhibits the perturbation acting twice to excite a two-photon transition from $|i\rangle$ to $|f'\rangle$ via all accessible intermediate states $|f\rangle$. To first order, $a_i^{(0)}(t) = 1 - O(V^2)$ and

$$a_f^{(1)}(t) = -\frac{i}{\hbar}\int_0^t dt' \, V_{fi}(t')e^{i\omega_{fi}t'}. \tag{3.21}$$

3.2 PERTURBATION THEORY AND THE GOLDEN RULE

Imagine that the applied classical radiation field which excites the atom is monochromatic and turned on suddenly at $t = 0$ with

$$\hat{V} = \hat{V}_0 \cos\omega t \tag{3.22}$$

where $\hat{V}_0 = -\hat{\underline{d}}.\underline{\mathcal{E}}_0$ and $\mathcal{E}_0$ is the slowly varying envelope of the electric field. We expand the cosine and integrate to give

$$a_f^{(1)}(t) = -\frac{1}{2\hbar}(V_0)_{fi}\left\{\frac{(e^{i(\omega+\omega_{fi})t}-1)}{(\omega+\omega_{fi})} - \frac{(e^{-i(\omega-\omega_{fi})t}-1)}{(\omega-\omega_{fi})}\right\}. \tag{3.23}$$

If the radiation field frequency ω is resonant with a transition frequency

ω_{fi}, the second term is much larger than the first, and non-resonant, background term. Dropping this "anti-resonant" term is equivalent to the "rotating-wave" approximation (e.g. Allen and Eberly 1975). If only the resonant term is kept we find the transition probability

$$P_f^{(1)}(t) = |a_f^{(1)}(t)|^2 = \frac{|(V_o)_{fi}|^2}{\hbar^2} \frac{\sin^2(\Delta t/2)}{\Delta^2} \tag{3.24}$$

where $\Delta = (\omega - \omega_{fi})$ is the detuning of the field from the atomic transition. For $\Delta = 0$, the maximum value of $P_f^{(1)}(t)$ is

$$P_f^{(1)}(\max) = |(V_o)_{fi}|^2/\hbar^2\Delta^2 \tag{3.25}$$

which, of course, has to be much less than one for the perturbation expansion to be valid. For $\Delta \approx 0$, we find $P_f^{(1)}(t) = |(V_o)_{fi}|^2 t^2/4\hbar^2$. However, this is only valid for short times t, since $P_f^{(1)}(t)$ is obliged to remain much less than unity. The evolution of the population in the excited state is shown in Fig. 3.1.

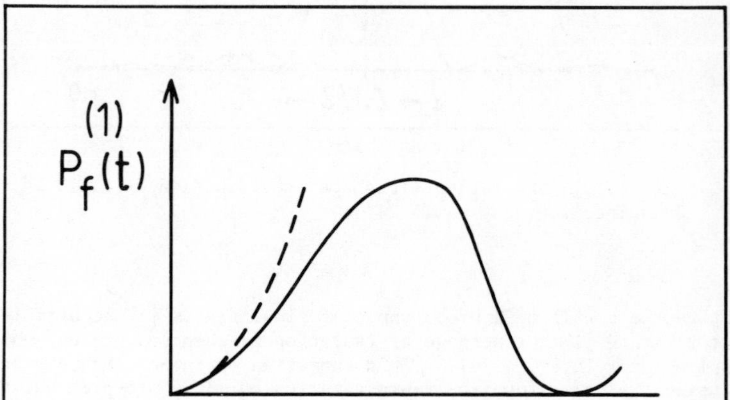

Fig. (3.1) The time development of the excited state population. (From eq.(3.24) for large detuning Δ, (solid line) and zero detuning $\Delta \approx 0$ (broken line)).

The periodic excitation and de-excitation shown in Figure 3.1 may alarm the reader who is familiar with the time-dependent transition rate called Fermi's Golden Rule. Under some circumstances we can find a time-independent transition rate which is valid for reasonably long times. The transition probability $P_f^{(1)}(t)$ given by eq.(3.24) is a sharply-peaked function shown in Fig.(3.2) with a width proportional to t^{-1} and a height proportional to t^2 so that the area is proportional to t. Mathematically

$$\frac{\sin^2 \Delta t/2}{\Delta^2} \xrightarrow[t\to\infty]{} \frac{\pi}{2} t\delta(\Delta). \tag{3.26}$$

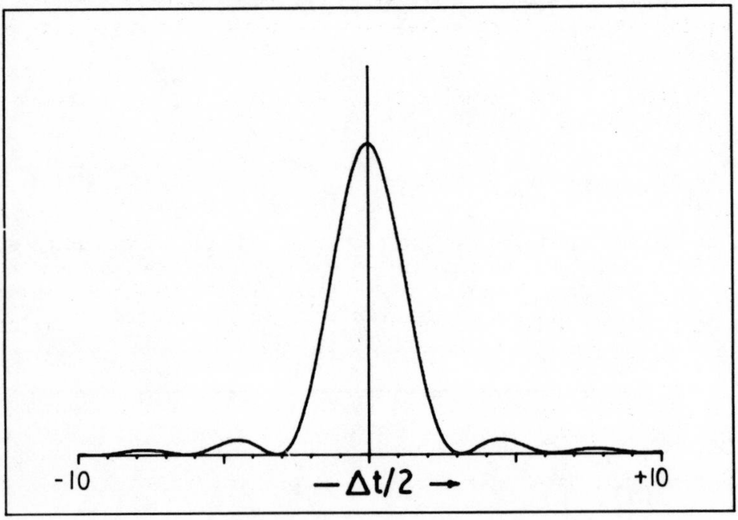

Fig. (3.2) Excitation lineshape as a function of field detuning Δ. (From eq.(3.24)).

In practice there will be a broad range of final states $|f\rangle$ accessible from the initial state $|i\rangle$, or a range of radiation frequencies present which have to be summed over in eq.(3.24). This summation or integration can be performed using the delta-function representation eq.(3.26) to pick out those energy-conserving states which dominate the excitation, to give

$$P_f^{(1)}(t) = \sum_f \frac{\pi}{2\hbar^2} |(V_o)_{fi}|^2 t\, \delta(\omega - \omega_{fi}) \tag{3.27}$$

and the time-independent rate $R = \dot{P}_f(t)$

$$R = \frac{\pi}{2\hbar^2} \sum_f |(V_o)_{fi}|^2\, \delta(\omega - \omega_{fi}). \tag{3.28}$$

This expression is sometimes called Fermi's Golden Rule and is usually written

$$R = \frac{2\pi}{\hbar^2} \sum_f |V_{fi}|^2\, \delta(\omega - \omega_{fi}) \tag{3.29}$$

where V_{fi} is the matrix element connecting f and i. In our calculation only the rotating part of eq.(3.22) is effective, which accounts for the factor of 4.

The long-time limit in eq.(3.26) and the resultant time-independent rate only makes sense when there is a broad band of energies contributing to the excitation. An example is light from a spectral lamp, where there is no phase relationship between the different frequency components of the light. The coupling $|(V_0)_{fi}|^2$ is then frequency dependent and the transition probability induced by all frequency components of the field is

$$P_f^{(1)}(t) = \frac{1}{\hbar^2}\int d\omega \, \frac{\sin^2\Delta t/2}{\Delta^2}\, I(\omega) \tag{3.30}$$

where

$$I(\omega) \equiv |(V_o(\omega))_{fi}|^2.$$

If $I(\omega)$ is so broad-band that it varies slowly compared with $(\sin^2\Delta t/2)/\Delta^2$, it may be replaced by its resonant peak value $I(\omega_{fi})$ and taken outside the integral. This can then be performed using eq.(3.26) to give

$$P_f^{(1)}(t) = \frac{\pi}{2\hbar^2}\, I(\omega_{fi})t \tag{3.31}$$

and the time-independent rate

$$R_f = \frac{\pi}{2\hbar^2}\, I(\omega_{fi}), \tag{3.32}$$

for the absorption of broad-band light. The washing-out of the oscillations visible in Fig.(3.1) is due to the spread of driving frequencies and the incoherence of the incident radiation. No such dephasing is present when the atom is driven by a coherent field and time-independent transition rates will not necessarily be appropriate.

3.3 THE RABI MODEL

The perturbation theory approach to the excitation of an atomic or molecular transition assumes that the initial state population is hardly changed by the excitation and that the probability amplitude for being in any other state remains small. A laser field which is near-resonant with a particular excited state will cause large population changes in that state, but not in others. In this case we must abandon perturbation theory, pick out the dominant state and try to solve the equations more exactly. In many cases only two states are important, (Fig.3.3) and we can expand the wavefunction as

$$\psi(t) = a_i e^{-iE_i t/\hbar}|i\rangle + a_f e^{-iE_f t/\hbar}|f\rangle \tag{3.33}$$

The laser perturbation which couples $|i\rangle$ and $|f\rangle$ is

$$\hat{V} = \hat{V}_o \cos\omega t \tag{3.34}$$

and we write

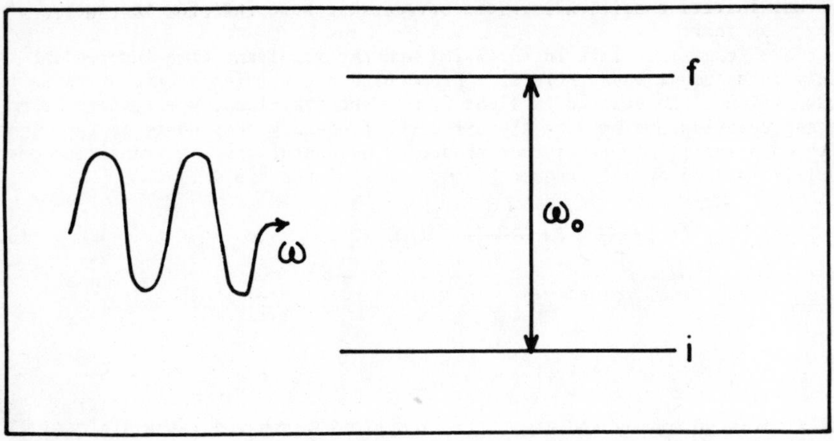

Fig. (3.3) Two-level atom excited by a monochromatic field

$$(V_o)_{fi} = \langle f|\hat{V}_o|i\rangle = (V_o)^*_{if} \tag{3.35}$$

From our general equation of motion for probability amplitudes, eq.(3.11), using only these two states we find

$$\dot{a}_i = -\frac{i}{\hbar}(V_o)_{if}\cos\omega t\ e^{-i\omega_o t}\ a_f \tag{3.36}$$

$$\dot{a}_f = -\frac{i}{\hbar}(V_o)_{if}\cos\omega t\ e^{i\omega_o t}\ a_i \tag{3.37}$$

where $\omega_o = (E_f - E_i)/\hbar$.

We will maintain the rotating-wave approximation; to do this we assume that $\omega \approx \omega_o$ and that exponents like $(\omega - \omega_o)$ will cause large resonant perturbations whereas the counter-rotating term in the exponent $(\omega + \omega_o)$ will oscillate rapidly and lead to little net effect. Dropping such terms gives

$$\dot{a}_i = -\frac{i}{2\hbar}(V_o)_{if}e^{i(\omega - \omega_o)t}\ a_f \tag{3.38}$$

$$\dot{a}_f = -\frac{i}{2\hbar}(V_o)_{if}e^{-i(\omega - \omega_o)t}\ a_i \tag{3.39}$$

These are the fundamental evolution equations which must be solved.

Our initial condition is given by starting with the atom in the ground state, so that

$$a_i(0) = 1 \ , \quad a_f(0) = 0 \tag{3.40}$$

and $\dot{a}_f(0) = -(i/2)(V_o)_{if}$; $\dot{a}_i(0) = 0$. We solve eqs.(3.38) and (3.39) by differentiation and substitution to give the second order equation in a_f alone,

$$\ddot{a}_f + i(\omega - \omega_o)\dot{a}_f + \frac{1}{4\hbar^2}|(V_o)_{if}|^2 a_f = 0 \ . \tag{3.41}$$

This has the trial solution

$$a_f(t) = \exp(i\mu t) \tag{3.42}$$

If we substitute eq.(3.42) into (3.41) we obtain a quadratic characteristic equation whose roots are

$$\mu_\pm = \frac{1}{2}\left\{-(\omega - \omega_o) \pm \sqrt{(\omega - \omega_o)^2 + |(V_o)_{if}|^2/\hbar^2}\right\} \tag{3.43}$$

so that

$$a_f(t) = A_+ e^{i\mu_+ t} + A_- e^{i\mu_- t} \tag{3.44}$$

where the initial conditions determine $A_\pm$. From (3.40) after a little algebra we find the Rabi solution for the excited state probability amplitude (e.g. Knight and Milonni 1980) to be,

$$a_f(t) = \frac{-(i/\hbar)(V_o)_{if}\, e^{-i(\omega - \omega_o)t/2}}{\sqrt{(\omega - \omega_o)^2 + |(V_o)_{if}|^2/\hbar^2}} \sin\left(\sqrt{(\omega - \omega_o)^2 + |(V_o)_{if}|^2/\hbar^2}\right)t/2 \ . \tag{3.45}$$

In Figure 3.4 we plot the probability of excitation, $|a_f(t)|^2$ as a function of time. The population is seen to vary periodically between the ground and excited states at the Rabi frequency, Ω.

$$\Omega = \sqrt{(\omega - \omega_o)^2 + |(V_o)_{if}|^2/\hbar^2} \ . \tag{3.46}$$

The maximum probability of excitation is

$$P_f(\max) = \frac{|(V_o)_{if}|^2/\hbar^2}{(\omega - \omega_o)^2 + |(V_o)_{if}|^2/\hbar^2} \tag{3.47}$$

and this can be unity, complete "population inversion", when the laser frequency ω is exactly on resonance with the atomic transition frequency ω_o; the atom then returns to the ground state and the cycle restarts. On exact resonance when $\omega = \omega_o$, a pulse of duration $t = \hbar\pi/|(V_o)_{if}|$ is capable of "inverting" the atom, and is called a π-pulse (see Allen and Eberly 1975 for an extensive discussion).

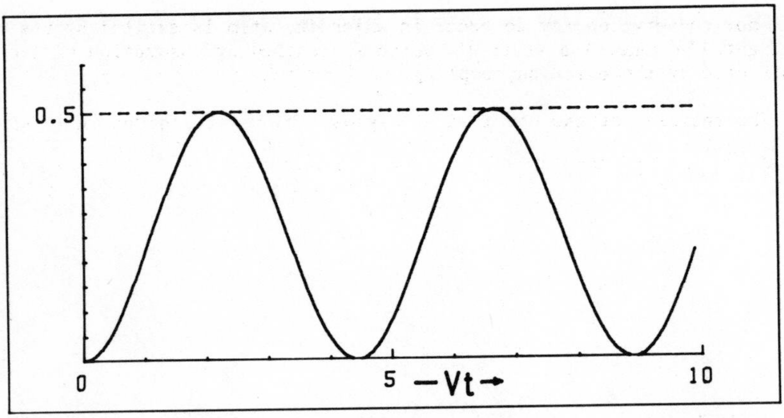

Fig. (3.4) The intense field Rabi solution for the time development of the excitation probability. (From eq.(3.45) for the case of $V/\Delta = 1$).

The perturbation theory results can be obtained either when the size of the perturbation $(V_0)_{if}$ is so weak compared with detunings $(\omega - \omega_0)$ that it can be neglected in Ω, or if the radiation acts for such a short time then the $\sin^2$ can be legitimately expanded. In both cases depletion becomes unimportant and the perturbation theory approach remains valid. The counter-rotating term we neglected leads to some intensity-dependent frequency shifts called Bloch-Seigert shifts (see Allen and Eberly 1975) which are essentially AC Stark shifts due to the interaction of the atom and the counter-rotating laser electric field.

3.4 EXCITATION OF AN ATOM BY A QUANTIZED RADIATION FIELD

So far we have considered the radiation to be entirely classical and exciting transitions between atomic states. In this section we return to the quantized description of the radiation field and show how a quantized field perturbation induces transitions between states of the combined atom plus field system. The initial state of the atom plus field is

$$|i\rangle = |a\rangle\,|n\rangle \tag{3.48}$$

where we consider the field to be a single-mode with n photons initially present, and the atom to be initially in state $|a\rangle$. The quantized perturbation can cause a transition to a state

$$|f_1\rangle = |b\rangle\,|n-1\rangle \tag{3.49}$$

with the absorption of a photon and to a state

$$|f_2\rangle = |b\rangle\,|n+1\rangle \tag{3.50}$$

by the emission of a photon. It is possible for a "virtual" transition which need not conserve energy to occur in which the atom is excited by the emission of light. In general a state $|b\rangle$ which is reached by absorption will be almost unaffected by the emission coupling.

The energies of the combined atom-plus-field states given in (3.48-50) are

$$\hbar\omega_i = E_a + n\hbar\omega \tag{3.51a}$$

and

$$\hbar\omega_{f_1} = E_b + (n-1)\hbar\omega \tag{3.51b}$$

$$\hbar\omega_{f_2} = E_b + (n+1)\hbar\omega. \tag{3.51c}$$

The perturbation is now time-independent. Using the form for the electric field operator $\hat{\underline{E}}$ discussed in Section (2.4) we have

$$\hat{V} = -\hat{\underline{d}}.\hat{\underline{E}} = -\hat{\underline{d}}.\underline{A}_o(\hat{a} - \hat{a}^+) \text{ where } \underline{A}_o = i(\hbar\omega/2\varepsilon_o V)^{\frac{1}{2}}\underline{\varepsilon} \tag{3.52}$$

This has matrix elements

$$\langle i|\hat{V}|f_1\rangle = V_{if_1} = +\underline{d}_{ab}.\underline{A}_o\sqrt{n} = V^*_{f_1 i} \tag{3.53}$$

where $\underline{d}_{ab} = \langle a|\hat{\underline{d}}|b\rangle$ and

$$\langle i|\hat{V}|f_2\rangle = V_{if_2} = -\underline{d}_{ab}.\underline{A}_o\sqrt{n+1} = V^*_{f_2 i}\ . \tag{3.54}$$

The first-order evolution equation (3.21) is still valid for the combined atom-plus-field system. The system is initially prepared in the state $|i\rangle$ and the interaction $\hat{V}$ turned on at $t = 0$. The probability amplitude for finding the atom in the excited state $|b\rangle$ irrespective of whether it reaches it from $|a\rangle$ by emission or absorption is

$$a_b^{(1)}(t) = -\frac{i}{\hbar}\int_0^t dt'\, V_{f_1 i}\, e^{i\omega_{f_1 i}t'} - \frac{i}{\hbar}\int_0^t dt'\, V_{f_2 i}\, e^{i\omega_{f_2 i}t'}\ . \tag{3.55}$$

Using eqs.(3.51) for the energies and eqs.(3.53) and (3.54) for the matrix elements gives

$$a_b^{(1)}(t) = +\frac{i}{\hbar}(\underline{d}_{ab}.\underline{A}_o)\left\{(n+1)^{\frac{1}{2}}\frac{(e^{i(\omega+\omega_{ba})t}-1)}{(\omega+\omega_{ba})} - n^{\frac{1}{2}}\frac{(e^{i(\omega-\omega_{ba})t}-1)}{(\omega-\omega_{ba})}\right\} \tag{3.56}$$

where the first term is due to emission, the second to absorption. If we recognise that for an intense incident field, $n \gg 1$ we can neglect the factor of unity in the first, emission term in eq.(3.56). Then the factor $n^{\frac{1}{2}}$ can be combined with the factors outside the curly brackets. We note the complete correspondence with the semiclassical result eq.(3.23) provided we identify $i(\underline{d}_{ab}\cdot\underline{A}_o)\sqrt{n}/\hbar$ with $-(V_o)_{fi}/2\hbar$. This identification is possible when we remember that the electric field is proportional to the square root of the field energy, and this in turn is proportional to $\sqrt{n}$, so that the electric field in eq.(3.22) can be equated to

$$\underline{\mathcal{E}}_o = 2i\underline{A}_o\sqrt{n} \tag{3.57}$$

The comments we made earlier concerning the rotating-wave approximation and the Golden Rule are equally applicable to the fully quantised result in eq. (3.56).

Note that a purely absorptive transition from $|a\rangle$ to $|b\rangle$ in rotating-wave approximation is described by the second term in eq.(3.56). Once the identification eq.(3.57) is made we recognise that there is no difference between semiclassical and fully-quantized treatments of weak field absorption. This remains true when the state $|b\rangle$ (or $|f\rangle$ in semiclassical theory) is replaced by a large number of accessible final states. Consequently a quantized field treatment is not necessary for a proper understanding of the photoelectric effect. A semiclassical discussion of which, based on first order perturbation theory and formulated in a similar manner to our discussion, is given in the Reprinted Paper 9 by Mandel, Sudarshan and Wolf. This paper also deals with fluctuations in the incident light field; a topic we shall approach in the next Chapter.

There are differences between semiclassical and full quantized theories of emission. Emission in eq.(3.56) is described by the first term and if $E_b < E_a$ (i.e. $|b\rangle$ is a lower-lying state than $|a\rangle$) emission is induced if $\omega \approx -\omega_{ba}$, and the second term in eq.(3.56) is now neglected in rotating-wave approximation. We note, from eq.(3.56) that emission persists even if $n = 0$ when the stimulating field is absent. This spontaneous emission is a purely quantum effect in this formalism. We refer the reader to standard textbooks (e.g. Loudon 1973; Sargent, Scully and Lamb 1974) for a derivation of Einstein A and B coefficients from the lowest-order perturbation theory result in eq.(3.55). But if spontaneous emission is unimportant there appear to be no differences between semiclassical and fully quantized treatments of radiative interactions. Differences occur whenever the operator nature of the field is important, and radiative corrections due to vacuum fluctuations and to spontaneous decay arise. Otherwise there is no need, or attraction, to invoke the photon concept in radiative interactions.

3.5 INDUCED DIPOLE MOMENTS

In our treatment of atomic excitation so far, we have assumed that the atomic energy levels are perfectly sharp. This is never the case because excited states decay by spontaneous emission and collisions de-excite levels. The uncertainty principle associates a finite energy width with the finite lifetime of the state. A useful way to model damping phenomenologically is to assume that the excited state probability for any state j decays at a rate $\gamma_j = 1/\tau_j$ where τ_j is the lifetime of the state. To achieve this we write the wave-function as

$$\psi_j(t) \approx e^{-iE_jt/\hbar}\phi_j \rightarrow e^{-iE_jt/\hbar - \gamma_jt/2}\phi_j \tag{3.58}$$

and the effect of making the energy of state j complex is to include the damping. Then from the general wave-function expansion, eq.(3.8)

$$\psi(t) = \sum_n a_n(t) e^{-iE_n t/\hbar} |n> \tag{3.59}$$

the probability of being in state $|n>$ at time t is

$$P_n(t) = |a_n(t) e^{-iE_n t/\hbar}|^2 = |a_n(t)|^2 e^{-\gamma_n t} \tag{3.60}$$

which reflects the well-verified exponential decay law of unstable states. To prove this ansatz correct from first principles would require extensive use of the quantum theory of damping (e.g. Heitler 1954; Haken 1969) and is clearly beyond the scope of this book. Nevertheless the outcome of such an analysis fully justifies the approach.

The basic equations of motion (eq.(3.11) for the atomic probability amplitudes can be easily modified to take decay into account. Let us consider the excitation of state $|f>$ by one photon absorption from the ground state (Fig.(3.5)).

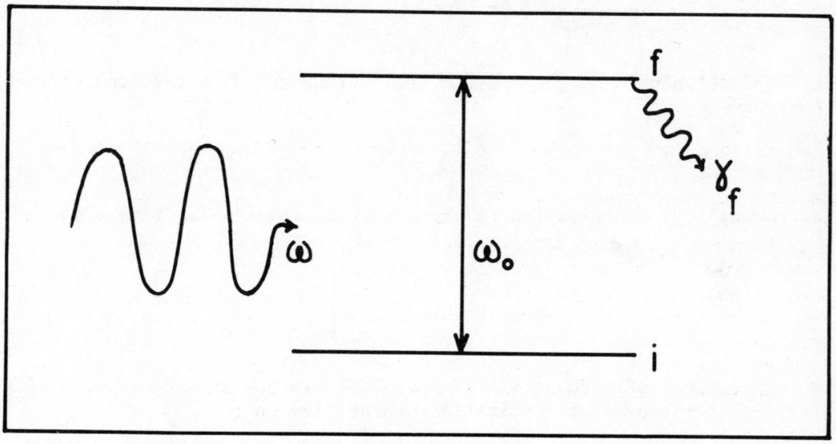

Fig. (3.5) Excitation of a decaying state by a monochromatic field.

We take the energy of the ground state to be zero and the excited state energy to be complex $E_f/\hbar = \omega_0 - i\gamma_f/2$. We consider the atom to be excited by radiation which is turned on suddenly at $t = 0$. The equation of motion for the excited state probability amplitude, from eqs.(3.36) and (3.37) is

$$\dot{a}_f = -\frac{i}{\hbar} V_{fi}(t) \exp[i(\omega_0 - i\gamma_f/2)t] a_i \tag{3.61}$$

Similarly we find for the ground state amplitude,

$$\dot{a}_i = -\frac{i}{\hbar}V_{if}(t)e^{-i(\omega_0 - i\gamma_f/2)t}a_f \ . \qquad (3.62)$$

As before, we imagine the incident field $\underline{E}(t)$ to be perfectly monochromatic

$$\hat{V}(t) = \tfrac{1}{2}\hat{V}_0(e^{-i\omega t} + e^{+i\omega t}) \ .$$

Again we make the rotating wave approximation and discard terms which oscillate rapidly at $\pm(\omega_0 + \omega)$ in eqs.(3.61) and (3.62). We then obtain

$$\dot{a}_f = -\frac{i}{2\hbar}(V_0)_{fi}\, e^{-i(\omega-\omega_0)t}\, e^{\gamma_f t/2}\, a_i \qquad (3.63)$$

$$\dot{a}_i = -\frac{i}{2\hbar}(V_0)_{if}\, e^{i(\omega-\omega_0)t}\, e^{-\gamma_f t/2}\, a_f \ . \qquad (3.64)$$

Again we solve these evolution equations in lowest order by assuming $a_1(t=0) = 1$; $a_1(t) \approx 1$. The excited state equation of motion eq.(3.63) can be directly integrated as in eq.(3.18) to give

$$a_f(t) = \frac{(i/2\hbar)(V_0)_{fi}}{i\Delta + \gamma_f/2}\left(e^{(-i\Delta + \gamma_f t/2)} - 1\right) \qquad (3.65)$$

where the detuning Δ of the applied field frequency from the atomic transition frequency is

$$\Delta = \omega - \omega_0 \ .$$

The probability of excitation to the unstable state $|f\rangle$ at time t is

$$P_f(t) = |a_f(t)|^2 e^{-\gamma_f t} = \frac{|(V_0)_{fi}|^2/4\hbar^2}{\Delta^2 + \gamma_f^2/4}\left[1 + e^{-\gamma_f t} - 2\cos\Delta t\, e^{-\gamma_f t/2}\right] . \qquad (3.66)$$

The oscillations at frequency Δ in eq.(3.24) are now damped by decay, Fig. (3.6). The steady-state excitation probability is

$$P_f(t \gg \gamma_f^{-1}) = \frac{|(V_0)_{fi}|^2/4\hbar^2}{(\Delta^2 + \gamma_f^2/4)} \qquad (3.67)$$

and is shown in Fig. 3.6. The excitation probability as a function of detuning is made up of an overall Lorentzian factor of width $\gamma_f = 1/\tau_f$ and a "ringing" factor given by the term in square brackets in eq.(3.66). For short times t, this factor dominates and the lineshape has a width t^{-1}; but for larger times the ringing decays away to leave a naturally-broadened steady-state line of width γ_f.

From eq.(3.65) for the excited state amplitude we can derive an expression for the dipole moment induced in the atom. This dipole is itself a source of radiation which interferes with the incident field to give rise to absorption and to the dispersive refractive index (e.g. Feynman, Leighton and Sands 1965;

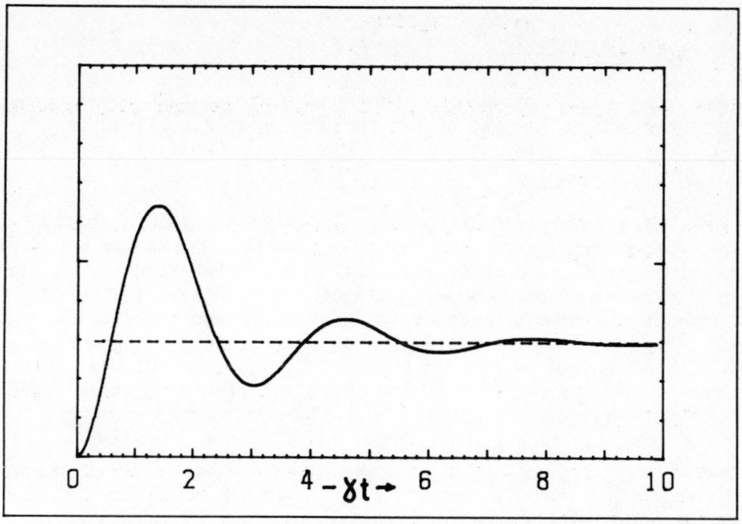

Fig. (3.6) The time development of an unstable state population excited by a pulse of coherent light. (From eq.(3.66) for the case of $\Delta/\gamma = 2$, $V/\gamma = 1$).

Sargent, Scully and Lamb 1974; Durrant, Reprinted Paper 10). For atoms initially in their excited state however, the dipoles give rise to a field which constructively interferes in the forward direction to give amplification rather than absorption of the incident field.

Imagine an incident field

$$\underline{E}^{(i)}(\underline{r},t) = \underline{\varepsilon}\, E_o \cos(\omega t - \underline{k}.\underline{r}) \tag{3.68}$$

where $\underline{\varepsilon}$ is the unit polarization vector. This field interacts with an atom at position $\underline{r}'$ at time t through the coupling

$$\hat{V} = -\,\hat{\underline{d}}.\underline{E}^{(i)}(\underline{r}',t). \tag{3.69}$$

At time t, the wavefunction of our two-level system made up of states $|f\rangle$ and $|i\rangle$ is, from eqs.(3.58), (3.59) and (3.65)

$$\psi(t) = |i\rangle +$$

$$+ \frac{(1/2\hbar)E_o d_{fi} e^{i\underline{k}.\underline{r}'}}{\omega_o - \omega - i\gamma_f/2}\left(e^{i(\omega_o-\omega)t + \gamma_f t/2} - 1\right) e^{-i\omega_o t}\, e^{-\gamma_f t/2}\, |f\rangle\ . \tag{3.70}$$

For times long enough for the transients to have died away ($t \gg \gamma_f^{-1}$), the steady-state superposition state is

$$\psi(t) = |i\rangle + \frac{E_o d_{fi}/2\hbar}{\Delta - i\gamma_f/2} e^{-i(\omega t - \underline{k}.\underline{r}')}|f\rangle . \tag{3.71}$$

The dipole moment induced by the incident radiation field is

$$\underline{D}(t) = \langle\psi(t)|\hat{\underline{d}}|\psi(t)\rangle . \tag{3.72}$$

If the atom were entirely excited or entirely in its ground state, $\underline{D}(t)$ would be zero. Under these circumstances the semi-classical theory would predict that the atom could not radiate and that a wholly inverted, or excited, atom would be entirely stable. However, the fully quantised theory reveals that the quantum dipole moment fluctuates, and the fluctuations would trigger off the decay (e.g. Milonni 1976, and references therein). A coherent superposition state such as that given by eq.(3.71) does however give a nonzero dipole moment. Using eqs.(3.71) and (3.72) we find after some algebra

$$D(t) = \frac{|d_{fi}|^2 E_o/\hbar}{\sqrt{\Delta^2 + \gamma_f^2/4}} \cos(\omega t - \underline{k}.\underline{r}' - \alpha) \tag{3.73}$$

where the phase lag α is given by

$$\tan\alpha = (\gamma_f/2)/\Delta . \tag{3.74}$$

The steady-state dipole moment oscillates at the incident field driving frequency ω, but lags in phase by α. On resonance, $\Delta = 0$ and $\alpha = \pi/2$; that is the phase of the induced moment lags the applied field by $\pi/2$.

The electric field at some position $\underline{r}$ a distance $R \gg \lambda$ from a dipole $\underline{D}(t)$ at $\underline{r}'$ is given by

$$\underline{E}(\underline{r},t) = (\omega^2/4\pi\varepsilon_o Rc^2)[\underline{D}(t - R/c) - \underline{n}.\underline{D}(t - R/c)\underline{n}] \tag{3.75}$$

where the unit vector $\underline{n}$ is

$$\underline{n} = (\underline{r} - \underline{r}')/|\underline{r} - \underline{r}'|$$

and $|\underline{r} - \underline{r}'| = R$ and $\underline{D}(t) = D(t)\underline{\varepsilon}$. Then using eq.(3.73) we find for the field radiated by the atomic dipole

$$\underline{E}(\underline{r},t) = (\omega^3/4\pi\varepsilon_o Rc^2)E_o[\underline{\varepsilon} - (\underline{n}.\underline{\varepsilon})\underline{n}] \frac{|d_{fi}|^2/\hbar}{\sqrt{\Delta^2 + \gamma_f^2/4}} \cos(\omega t - \underline{k}.\underline{r}' - \alpha - kR) . \tag{3.76}$$

The field radiated from the atom in its coherent superposition state has a dipole angular pattern. In the forward $\underline{k}$ direction $\underline{n}.\underline{\varepsilon} = 0$ and $\underline{k}.\underline{r}' + kR = \underline{k}.\underline{r}$ we find for the radiated field

$$\underline{E}(\underline{r},t) = (\omega^2/4\pi\varepsilon_o Rc^2) \frac{E_o \underline{\varepsilon}|d_{fi}|^2/\hbar}{\sqrt{\Delta^2 + \gamma_f^2}} \cos(\omega t - \underline{k}.\underline{r} - \alpha) . \tag{3.77}$$

On resonance the phase-lag α is $\pi/2$ and the forward dipole field lags the incident field by $\pi/2$. The field radiated by many dipoles, however, does not phase-lag the incident field. The properly-phased vector-addition of the fields produced by many atoms in a sheet of polarizable medium produces an additional phase-lag of $\pi/2$ in the forward direction (see Reprinted paper by Durrant), so that the resultant total forward field is out of phase with the incident field. An almost identical argument for initially inverted atoms sheds light on the spatial properties of stimulated emission, but with the result that the forward amplified field is in-phase with the incident field.

The calculation of the dipole moment in this section has used lowest order perturbation theory. The intense-field Rabi solution can be used to construct a non-perturbative expression for the dipole moment which is modulated by the periodic Rabi oscillations. Such an oscillating moment would give a time-dependent emission signal called "optical nutation" (see Allen and Eberly 1975). Fourier resolution would reveal sidebands in the spectrum separated by the Rabi frequency from the centre component which appears at the driving frequency ω. As $t \to \infty$ these Rabi oscillation transients die away and the strongly-driven dipole moment again oscillates only at ω. Remarkably, the sidebands are observed to persist even in steady-state (Schuda et al 1974, Reprinted paper 11). This is interpreted as due to quantum fluctuations in the dipole evolution which re-initiate the transients and maintain the sidebands (see for example Knight and Milonni 1980 and references therein).

CHAPTER 4

Coherence Functions

4.1 CLASSICAL COHERENCE

In this Chapter we show how the classical notions of optical coherence can be generalized to describe correlations in quantized electromagnetic fields. We begin by reviewing elementary notions of coherence, show how higher order correlations can be described in classical theory and present an analysis of quantum coherence based on the analysis of Glauber. It is surprising how much of the classical formalism is found to remain useful.

We assume that the reader is familiar with the basic notions of classical coherence (e.g. Fowles, 1975; Hecht and Zajac 1974). In Young's two-slit experiment it is easy to demonstrate that the intensity at a point on the screen where the interference fringes are observed, is given in terms of the fields E_1 and E_2 at the slits 1 and 2 by

$$I = \langle |E_1|^2 \rangle + \langle |E_2|^2 \rangle + 2 \operatorname{Re}\langle E_1 E_2^* \rangle \tag{4.1}$$

If the difference in time for the light to travel to the observation point from the slits is τ, then the interference term $2 \operatorname{Re}\langle E_1 E_2^* \rangle$ may be re-written using the correlation function, or mutual coherence function, (Born and Wolf 1964; Mandel and Wolf 1965) as

$$\Gamma_{12}(\tau) = \langle E_1(t) E_2^*(t+\tau) \rangle. \tag{4.2}$$

The normalized form of the correlation function, or the degree of partial coherence, is

$$\gamma_{12}(\tau) = \frac{\Gamma_{12}(\tau)}{\sqrt{\Gamma_{11}(0)\ \Gamma_{22}(0)}} = \frac{\Gamma_{12}(\tau)}{\sqrt{I_1\ I_2}} \tag{4.3}$$

and we see that

$$I = I_1 + I_2 + 2\sqrt{I_1 I_2}\ \operatorname{Re} \gamma_{12}(\tau) \tag{4.4}$$

If Rayleigh's definition of fringe visibility is now used

$$V = (I_{max} - I_{min})/(I_{max} + I_{min})$$

we see at once that $V = 2\sqrt{I_1 I_2}\ |\gamma_{12}|/(I_1 + I_2)$ because

$$I_{\substack{max\\min}} = I_1 + I_2 \pm 2\sqrt{I_1 I_2}\,|\gamma_{12}|.$$

For equally illuminated slits this yields the physically identifiable result

$$V = |\gamma_{12}| \;;$$

in other words, the degree of fringe contrast as measured by the visibility is a direct measure of $|\gamma_{12}|$.

Glauber (1965) showed how the classical correlation function $\langle E_1(t)E_2^*(t+\tau)\rangle$ could be generalized to all orders in such a way that their quantized field equivalents followed in a natural and understandable manner. The electric, and magnetic fields involved in the system of interest can be expanded into modes by assuming there to be a cavity of appropriate volume V as in Chapter 2. The mode functions appropriate to a box of side-length L made up of perfectly conducting walls are, for example (Power 1964; Loudon 1973; Glauber 1965, 1970)

$$\underline{u}_k(r) = L^{-3/2}\,\underline{\varepsilon}_\lambda\, e^{i\underline{k}.\underline{r}} \tag{4.5}$$

where $\underline{\varepsilon}_\lambda$ ($\lambda = 1,2$) is the unit polarization vector. The multi-mode cavity field is determined by a product of these mode-functions and time-varying frequency factors and can be usefully decomposed into positive and negative frequency parts.

$$\underline{E}(\underline{r},t) = \underline{E}^{(+)}(\underline{r},t) + \underline{E}^{(-)}(\underline{r},t) \tag{4.6}$$

where $\underline{E}^{(-)}(\underline{r},t) = \underline{E}^{(+)}(\underline{r},t)^*$ and

$$\underline{E}^{(+)}(\underline{r},t) = \sum_k C_k\, \underline{u}_k(r)e^{-i\omega_k t}\;. \tag{4.7}$$

These positive and negative frequency parts of the field are related to the analytic signal of semi-classical theory discussed by Mandel, Sudarshan and Wolf [Reprinted Paper 9]. The expansion coefficients C_k are in general random variables (Glauber 1965) with a probability distribution $p(\{C_k\}) = p(C_1,C_2\ldots)$ which is normalized

$$\int P(\{C_k\})\;\Pi_k\; d^2C_k = 1 \tag{4.8}$$

where the differential element is $d^2C_k \equiv d(\mathrm{Re}\;C_k)d(\mathrm{Im}\;C_k)$. If we neglect the unimportant vector nature of the field, the average intensity may be shown (e.g. Loudon 1973; Glauber 1965) to be,

$$\langle|E^{(+)}(\underline{r},t)|^2\rangle = \langle E^{(-)}(\underline{r},t)E^{(+)}(\underline{r},t)\rangle \tag{4.9}$$

where the average is over the expansion coefficients $\{C_k\}$.

In general the average of a function of $E^{(+)}$ is

$$\langle F(E^{(+)}(\underline{r},t)\rangle = \int P(\{C_k\})F[E^{(+)}(\underline{r},t,\;\{C_k\})]\;\Pi_k\;d^2C_k \tag{4.10}$$

The first order correlation function may be used to demonstrate an example of the use of this. The classical first order correlation function is defined as

$$G(\underline{r}\ t;\ \underline{r}'\ t') = \langle E^{(-)}(\underline{r},t)E^{(+)}(\underline{r}',t')\rangle$$

and becomes

$$G = \int P(\{C_k\})E^{(-)}(\underline{r},t,\{C_k\})E^{(+)}(\underline{r}',t',\{C_k\})\ \Pi_k\ d^2C_k \tag{4.11}$$

4.2 QUANTUM COHERENCE FUNCTIONS

Glauber in a series of papers and articles in the 1960's showed how the quantum theory of coherence can be constructed in a manner which emphasizes observables and closely follows the readily accessible classical theory of coherence. The positive frequency part of the quantized multimode electric field operator may be expanded as

$$\hat{\underline{E}}^{(+)}(\underline{r},t) = i\sum_k \left(\frac{\hbar\omega_k}{2}\right)^{\frac{1}{2}} \hat{a}_k\ \underline{u}_k(\underline{r})e^{-i\omega_k t}\ . \tag{4.12}$$

We will be concerned with the description of fluctuations in the light field and their detection, and need first to describe the action of an ideal photon counter (Glauber 1965, 1970; Loudon 1973; Nussenzweig 1973). An ideal photon counter detects radiation by absorption. We consider our ideal detector to be a single atom of dimension small compared with the wavelength of the light. The detector couples to the quantized light field by a dipole interaction

$$\hat{V} = -\ \hat{\underline{d}}.\hat{\underline{E}}(\underline{r},t). \tag{4.13}$$

From eq.(3.18) generalised to an arbitrary initial and final state of the radiation field, we may determine the transition amplitude to go from the initial state $|i\rangle$ of the field to the final state $|f\rangle$ through absorption. We want to describe the quantized field in the interaction representation (Matthews 1974) in which field operators have a time dependence dictated by the unperturbed free-field Hamiltonian $\hat{\mathcal{H}}_F$, so that at time t,

$$\hat{O}(t) = e^{i\hat{\mathcal{H}}_F t/\hbar}\ \hat{O}\ e^{-i\hat{\mathcal{H}}_F t/\hbar}$$

where $\hat{O}$ is the Schrödinger picture operator. At $t = 0$, the interaction and Schrödinger pictures coincide. The advantage of the interaction picture is that the coupling between atom and field is now time-dependent exactly as in the semi-classical theory, and that free-field Hamiltonians do not explicitly appear in the dynamics (e.g. Pegg 1980). The fully quantized Hamiltonian in the interaction representation is

$$\hat{\mathcal{H}} = \hat{\mathcal{H}}_d - \hat{\underline{d}}(t).\hat{\underline{E}}(t) \tag{4.14}$$

where $\hat{\mathcal{H}}_d$ is the detector Hamiltonian and $\hat{\underline{E}}(t) = e^{-\hat{\mathcal{H}}_F t/\hbar}\ \hat{\underline{E}}e^{i\hat{\mathcal{H}}_F t/\hbar}$. As $\hat{\mathcal{H}}_F$ no longer appears in eq.(4.10), the unperturbed field energies will not appear in equations of motion for probability amplitudes; however, this formulation does display the field time-dependence explicitly. We imagine the detector responds to the radiation by absorption, making a transition from the initial state $|g\rangle$ with the field in state $|\psi_i\rangle$ to the final state $|e\rangle$ with the field in state $|\psi_f\rangle$, so

$$\psi(t) = c_i|g, \psi_i\rangle + c_f|e, \psi_f\rangle. \tag{4.15}$$

The time-dependent Schrödinger equation

$$i\hbar \frac{\partial}{\partial t} \psi(t) = (\hat{\mathcal{H}}_d - \hat{\underline{d}}.\hat{\underline{E}}(t))|\psi(t)\rangle$$

then generates the equations of motion for the amplitudes $c_i(t) = a_i(t)e^{-i\omega_i t}$, which can be integrated in lowest order perturbation theory as in eq.(3.18) to give,

$$a_e(t) = +\frac{i}{\hbar}\int_0^t dt'\ \underline{d}_{eg} e^{i\omega_{eg}t'} \cdot \langle\psi_f|\hat{\underline{E}}(t')|\psi_i\rangle \tag{4.16}$$

where $\underline{d}_{eg} = \langle e|\hat{\underline{d}}|g\rangle$ is the transition dipole moment. The total electric field operator is $\hat{\underline{E}}(t') = \hat{\underline{E}}^{(+)}(t') + \hat{\underline{E}}^{(-)}(t')$ and within the rotating-wave approximation only $\hat{\underline{E}}^{(+)}$ plays a role in absorption. We neglect the vector nature of the field from now on.

The probability of the detector being excited to state $|e\rangle$ is then $|a_e(t)|^2$, which using eq.(4.16) is

$$P_{ef}(t) = |d_{eg}|^2 \int_0^t dt' \int_0^t dt''\ e^{i\omega_{eg}(t''-t')}$$

$$\times\ \langle\psi_i|\hat{E}^{(-)}(t')|\psi_f\rangle\langle\psi_f|\hat{E}^{(+)}(t'')|\psi_i\rangle\ \hbar^{-2}\ . \tag{4.17}$$

We are really only interested in the final state of the detector and not that of the field, so we must sum eq.(4.17) over all possible final states $|\psi_f\rangle$ accessible from $|\psi_i\rangle$ by one-photon absorption. Since the states are a complete set,

$$\sum_f |\psi_f\rangle\langle\psi_f| = 1$$

so that the probability of the detector being excited irrespective of the final state of the field can be written as

$$P_e(t) = |d_{eg}|^2 \int_0^t dt' \int_0^t dt''\ e^{i\omega_{eg}(t''-t')}\langle\psi_i|\hat{E}^{(-)}(t')\hat{E}^{(+)}(t'')|\psi_i\rangle\hbar^{-2}. \tag{4.18}$$

We see immediately that the excitation is determined by a field correlation function.

The two time integrals in eq.(4.18) can be removed if we consider an idealised broadband detector for our attention. Many final states with a range of energies much wider than the bandwidth of the radiation field contribute to the absorption otherwise the detector would only see certain Fourier components and not respond to others; the energy range is also much wider than the reciprocal of the detection time, t^{-1}. The probability of the detector's absorbing a photon and becoming excited, irrespective of the particular final state of the field or detector is

$$P(t) = \int \rho(\omega_{eg}) P_e(t) d\omega_{eg} \qquad (4.19)$$

where $\rho(\omega_{eg})$ is the density of final excited states of the detector. As we may safely extend the integration over ω_{eg} to all frequencies, and $\rho(\omega_{eg})$ is slowly varying and can be replaced by a constant value,

$$\int_{-\infty}^{\infty} d\omega_{eg} e^{i\omega_{eg}(t''-t')} = 2\pi\delta(t''-t')\rho(\bar{\omega}_{eg}) \ . \qquad (4.20)$$

The detector excitation probability becomes

$$P(t) = S \int_{0}^{t} dt' \langle\psi_i|\hat{E}^{(-)}(t')\hat{E}^{(+)}(t')|\psi_i\rangle \qquad (4.21)$$

where S contains all of the detector constants. The rate of detector excitation, or counting rate, is

$$W(t) = \frac{dP}{dt} = S\langle\psi_i|\hat{E}^{(-)}(t)\hat{E}^{(+)}(t)|\psi_i\rangle \qquad (4.22)$$

In other words our ideal photodetector responds to the instantaneous local field intensity.

Probability amplitudes such as $c_e(t)$ are not directly measurable. Instead we measure bilinear combinations of amplitudes such as probabilities $|c_e(t)|^2$ or coherences $c_e^*(t)c_g(t)$. It is convenient to work directly with these bilinear quantities, and this may be accomplished using the density matrix [ter Haar 1961]. Imagine a two-state system with

$$|\psi\rangle = c_1|1\rangle + c_2|2\rangle \ . \qquad (4.23)$$

It is useful to form the products

$$\begin{aligned} \rho_{11} &= c_1 c_1^* \\ \rho_{12} &= c_1 c_2^* \\ \rho_{21} &= c_2 c_1^* = \rho_{12}^* \\ \rho_{22} &= c_2 c_2^* \end{aligned} \qquad (4.24)$$

or in matrix form

$$\hat{\rho} = \begin{pmatrix} \rho_{11} & \rho_{12} \\ \rho_{21} & \rho_{22} \end{pmatrix} \equiv |\psi\rangle\langle\psi| \qquad (4.25)$$

which is the density matrix for the two-state system. It contains all of the measurable information about our system.

The expectation value of an operator $\hat{O}$ can be written as $\langle\hat{O}\rangle = \langle\psi|\hat{O}|\psi\rangle$ or in our new notation as,

$$\langle\hat{O}\rangle = \rho_{11}O_{11} + \rho_{12}O_{21} + \rho_{21}O_{12} + \rho_{22}O_{22}$$

$$= \sum_i \left[\sum_j \rho_{ij}O_{ji}\right] = \sum_i (\hat{\rho}\hat{O})_{ii} = \mathrm{Tr}(\hat{\rho}\hat{O}) \ . \qquad (4.26)$$

where Tr denotes the trace; that is, the sum of the diagonal elements.

We can therefore write our photon counting rate eq.(4.22) as

$$W(t) = S\,\mathrm{Tr}\{\rho\hat{E}^{(-)}(t)\hat{E}^{(+)}(t)\} \ . \qquad (4.27)$$

Sometimes we cannot find the appropriate set of wavefunctions to describe our system, and all we have is a set of statistical probabilities that the system was prepared in various possible states (e.g. Loudon 1973). In this situation we must generalise the density operator from the "pure" case of eq.(4.25) to that for a statistical mixture. For a particular realisation r of the ensemble, the expectation value is $\langle r|\hat{O}|r\rangle$. The ensemble average is therefore

$$\langle\hat{O}\rangle = \sum_r P_r\langle r|\hat{O}|r\rangle \qquad (4.28)$$

where P_r is a normalised probability, $\sum_r P_r = 1$, for the realisation r. The mixed-state density operator, from eqs.(4.25), (4.27) and (4.28), is then

$$\hat{\rho} = \sum_r P_r|r\rangle\langle r| \qquad (4.29)$$

with $\langle\hat{O}\rangle = \mathrm{Tr}(\hat{\rho}\hat{O})$ as before. For a pure state with a single realisation k, $P_k = 1$, $P_{r\neq k} = 0$ we find $\hat{\rho}$ reduces to $|k\rangle\langle k|$. We should stress that for a pure state

$$\hat{\rho}^2 = \hat{\rho} \qquad (4.30)$$

from eq.(4.25), but this is not true for a mixed state.

Using this density matrix formalism we have the detector response at position $\underline{r}$ in the form given by eq.(4.27). The detected field intensity, ignoring the uninteresting detector constant S, is then

$$I = \mathrm{Tr}[\hat{\rho}\hat{E}^{(-)}(\underline{r},t)\hat{E}^{(+)}(\underline{r},t)] = \mathrm{Tr}\,[\hat{\rho}\hat{I}] \qquad (4.31)$$

where $\hat{I}$ is the intensity operator. We note that the intensity operator is "normally ordered" with annihilation operators to the right of creation operators: this is a consequence of the absorptive nature of our detector. Such a detector does not register vacuum fluctuations, and if the field were originally in a multimode vacuum state with no photons in any mode, $\lambda_1 \ldots \lambda_n$,

$$\hat{\rho} = |\{0\}><\{0\}| \equiv |0_{\lambda_1}>|0_{\lambda_2}>----|0_{\lambda_n}><0_{\lambda_n}|---<0_{\lambda_2}|<0_{\lambda_1}| \ ,$$

then $I = 0$ simply because a detector cannot absorb from such a field.

The quantum average $I(\underline{r},t)$ defined in eq.(4.31) can be generalized by taking field operators at different space-time points ($x_1 = r_1\ t_1$; $x_2 = r_2\ t_2$) to define the first order coherence function

$$G^{(1)}(x_1;x_2) = \mathrm{Tr}[\rho\hat{E}^{(-)}(x_1)\hat{E}^{(+)}(x_2)] \tag{4.32}$$

and more generally the nth order Glauber coherence function

$$G^{(n)}(x_1;x_2;---x_{2n}) = \mathrm{Tr}[\hat{\rho}\hat{E}^{(-)}(x_1)---\hat{E}^{(-)}(x_n)\hat{E}^{(+)}(x_{n+1})---\hat{E}^{(+)}(x_{2n})]. \tag{4.33}$$

From eq.(4.31) we see that $G^{(1)}(x_1;x_1) = I(\underline{r},t)$.

4.3 THE YOUNG TWO-SLIT EXPERIMENT

The quantum theory of the Young's two-slit interference experiment provides a simple example of the use of the Glauber coherence functions. Consider the idealised wavefront interference shown in Fig.(4.1) with a light from two very small slits at $\underline{r}_1$ and $\underline{r}_2$ being detected at point $\underline{r}$ at the "screen". The operator field $\hat{E}^{(+)}(\underline{r},t)$ at the detection point is given, following conventional optics, as

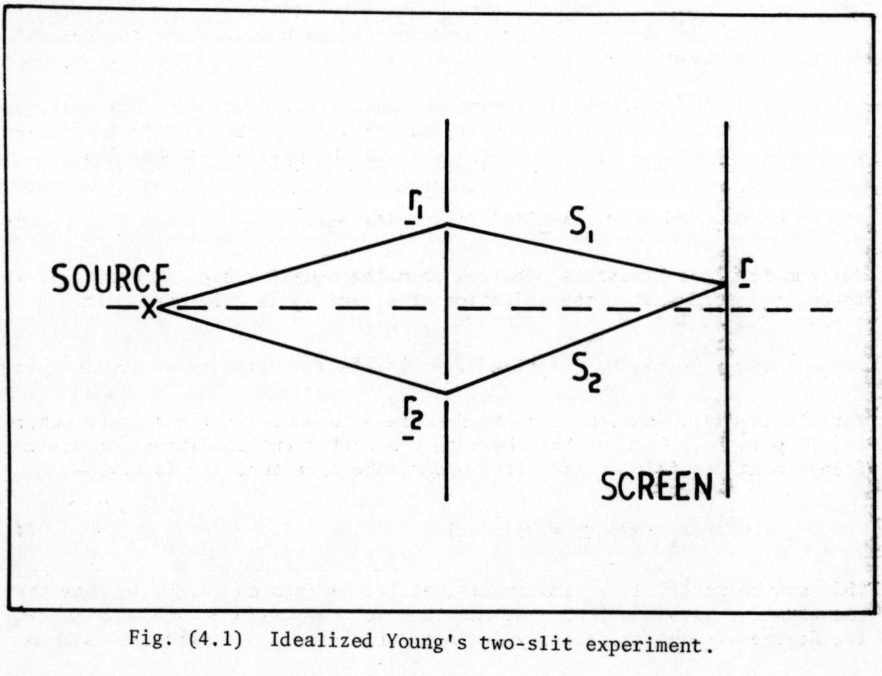

Fig. (4.1) Idealized Young's two-slit experiment.

$$\hat{E}^{(+)}(\underline{r},t) = \lambda[\hat{E}^{(+)}(\underline{r}_1, t - \frac{s_1}{c}) + \hat{E}^{(+)}(\underline{r}_2, t - \frac{s_2}{c})] \tag{4.34}$$

in other words this depends on the fields at the two slits at the appropriate times. This may then be written

$$\hat{E}^{+}(\underline{r},t) = \lambda[\hat{E}^{(+)}(x_1) + \hat{E}^{(+)}(x_2)]$$

where λ is a constant containing geometrical factors and x again denotes both space and time. It is useful to realise that the electric field operator $\hat{E}^{(+)}(x)$ obeys exactly the same Maxwell wave equation as its classical counterpart so that results from the classical theory of diffraction can be used directly in the quantum theory. The detected intensity at $(\underline{r},t)$ is, from eqs. (4.31) and (4.34)

$$I(\underline{r},t) = Tr[\hat{\rho}\hat{E}^{(-)}(\underline{r},t)\hat{E}^{(+)}(\underline{r},t)]$$

$$= |\lambda|^2[G^{(1)}(x_1; x_1) + G^{(1)}(x_2; x_2) + 2 \text{ Re } G^{(1)}(x_1; x_2)] . \tag{4.35}$$

We define a phase ϕ for each correlation function by writing,

$$G^{(1)}(x_i;x_j) = |G^{(1)}(x_i;x_j)|e^{i\phi} \tag{4.36}$$

so that

$$I(\underline{r},t) = |\lambda|^2\{G^{(1)}(x_1; x_1) + G^{(1)}(x_2; x_2) + 2|G^{(1)}(x_1; x_2)|\cos\phi\} \tag{4.37}$$

which may be compared with the classical equivalent eq. (4.4). If the detector is moved the phase ϕ changes, because the path difference changes, and fringes are produced.

The fringe contrast is determined by the magnitude of $|G^{(1)}(x_1; x_2)|$. The contrast is high if the fields at $\underline{r}_1$ and $\underline{r}_2$ are coherent but is limited by the requirement that $I(\underline{r},t)$ be positive, which implies (Glauber 1965)

$$G^{(1)}(x_1; x_1)G^{(1)}(x_2; x_2) \geq |G^{(1)}(x_1; x_2)|^2 . \tag{4.38}$$

Maximum fringe contrast is obtained when the equality sign in eq. (4.38) is taken, and occurs when the radiation at x_1 and x_2 is coherent. If

$$G^{(1)}(x_1; x_1)G^{(1)}(x_2; x_2) = |G^{(1)}(x_1; x_2)|^2 \tag{4.39}$$

for all $(x_1; x_2)$, then the radiation is said to be fully first order coherent at all points. Glauber has shown that a sufficient condition for maximum fringe contrast is that the first order coherence function factorizes

$$G^{(1)}(x_1; x_2) = \mathcal{E}^*(x_1)\,\mathcal{E}(x_2) . \tag{4.40}$$

This can be verified by substituting eq. (4.40) into eq. (4.38) to show that the equality is satisfied. We should note that we do not need to specify the degree of monochromaticity in this definition. Nevertheless such a specification is implicit in eq.(4.40) provided we restrict our attention to stationary fields, that is those invariant under time translation. Most fields of practical interest possess this property. If the radiation is stationary, the coherence function $G^{(1)}$ depends on the time difference $(t_1 - t_2)$ and not on times t_1 and t_2 separately.

$$G^{(1)}(\underline{r}_1\ t_1;\underline{r}_2\ t_2) = G^{(1)}(\underline{r}_1\ t_1+\tau;\underline{r}_2\ t_2+\tau)$$

$$= G^{(1)}(\underline{r}_1,\underline{r}_2;t_1-t_2) \tag{4.41}$$

for all translations τ. If the radiation has full first-order coherence then by eq.(4.40), $G^{(1)}(\underline{r}_1t_1;\underline{r}_2t_2) = G^{(1)}(\underline{r}_1,\underline{r}_2;t_1-t_2) = \mathcal{E}^*(\underline{r}_1,t_1)\,\mathcal{E}(\underline{r}_2,t_2)$ and this can be satisfied only by an exponential time-dependence

$$\mathcal{E}(t) = \mathcal{E}(0)e^{-i\omega t} \tag{4.42}$$

which implies that the first order coherent fields must be monochromatic. A generalization of the coherence factorization condition eq.(4.40) is readily possible: a field is fully nth order coherent if

$$G^{(m)}(x_1----x_{2m}) = \mathcal{E}^*(x_1)----\mathcal{E}^*(x_m)\,\mathcal{E}(x_{m+1})----\mathcal{E}(x_{2m}) \tag{4.43}$$

for all $m \le n$.

4.4 SECOND-ORDER CORRELATION FOR SINGLE-MODE RADIATION

As an example of the use of the quantum coherence functions we calculate the second-order coherence function $G^{(2)}$ where

$$G^{(2)}(\underline{r}_1t_1;\underline{r}_2t_2;\underline{r}_2t_2;\underline{r}_1t_1) = \langle\hat{E}^-(\underline{r}_1,t_1)\hat{E}^-(\underline{r}_2,t_2)\hat{E}^+(\underline{r}_2,t_2)\hat{E}^+(\underline{r}_1,t_1)\rangle. \tag{4.44}$$

Just as for the classical coherence function, eq.(4.3), we find it useful to normalise this

$$g^{(2)}(\underline{r}_1t_1;\underline{r}_2t_2;\underline{r}_2t_2;\underline{r}_1t_1) \equiv \frac{G^{(2)}(\underline{r}_1t_1;\underline{r}_2t_2;\underline{r}_2t_2;\underline{r}_1t_1)}{\langle\hat{E}^-(\underline{r}_1,t_1)\hat{E}^+(\underline{r}_1,t_1)\rangle\langle\hat{E}^-(\underline{r}_2,t_2)\hat{E}^+(\underline{r}_2,t_2)\rangle}. \tag{4.45}$$

For a single-mode field, the degree of second order coherence can be expressed in terms of the mean, and mean square, photon numbers. We restrict ourselves to the same point in space and time, $x_1 = x_2$, and find, using eqs.(4.12) and (4.44)

$$g^{(2)}(0) = \frac{\langle\hat{a}^+\hat{a}^+\hat{a}\ \hat{a}\rangle}{\langle\hat{a}^+\hat{a}\rangle^2} \tag{4.46}$$

This normalized form of $G^{(2)}$ is used in the Reprinted Paper 12 by Loudon to discuss the statistical properties of various kinds of single-mode radiation. It can be rewritten, using the basic commutation relation, $\hat{a}^+\hat{a} = \hat{a}\,\hat{a}^+ - 1$ and the number operator $\hat{n} = \hat{a}^+\hat{a}$ as

$$g^{(2)}(0) = \frac{\langle \hat{a}^+\hat{a}\,\hat{a}^+\hat{a}\rangle - \langle \hat{a}^+\hat{a}\rangle}{\langle \hat{a}^+\hat{a}\rangle^2} \tag{4.47}$$

The use of the variance $\sigma^2 = \langle \hat{n}^2\rangle - \langle \hat{n}\rangle^2$ enables us to write eq.(4.47) as

$$g^{(2)}(0) = 1 + \frac{(\sigma^2 - \langle \hat{n}\rangle)}{\langle \hat{n}\rangle^2} \tag{4.48}$$

as used in the Reprinted Paper 13 by Walls. For a number-state field, there is no uncertainty in the number of photons in the mode and the variance σ^2 is zero, and for such a state

$$\begin{aligned} g^{(2)}(0) &= 1 - 1/n \qquad & n \geq 2 \\ &= 0 \qquad & n < 2 \end{aligned} \tag{4.49}$$

which is always less than unity. We shall see in Chapter 6 that a field with $g^{(2)} < 1$ can be described only quantum mechanically.

CHAPTER 5

Coherent States

5.1 PHASE AND SUPERPOSITION

Number states are eigenstates of the field Hamiltonian and exhibit no phase information. For example, the electric field operator has zero expectation value in a number state

$$<n|\hat{\underline{E}}|n> = <n|(\hat{\underline{E}}^{(+)} + \hat{\underline{E}}^{(-)})|n> = 0 \tag{5.1}$$

simply because $\underline{E}^{(\pm)}$ creates (annihilates) photons, to produce $|n \pm 1>$, and orthogonality gives $<n|n \pm 1> = 0$. The mean square electric field $<n|\hat{\underline{E}}^2|n>$ is non-zero, but is proportional to the field energy $n\hbar\omega$. We can visualize this field as if its phase were uniformly distributed over 2π so that its mean is zero but its mean square is finite. (Loudon 1973). Phase or coherence is exhibited in general only by a superposition of quantum energy eigenstates. For example, if we were to try a superposition of number states differing by ±1 such as

$$\psi = c_n|n> + c_{n'}|n'> \tag{5.2}$$

where $n' = n \pm 1$, then this state would give a non-zero mean electric field $<\psi|\hat{\underline{E}}|\psi>$.

Optics is concerned almost entirely with coherence. A fully quantized description of optical coherence is naturally dominated by superposition states such as eq.(5.2). A fully coherent light field would be described by a superposition of number states with a large spread in photon numbers. Such a superposition is the coherent state $|\alpha>$ which we introduce in this chapter. We will show that $|\alpha>$ can be expressed as a Poisson superposition of number states,

$$|\alpha> = \sum_n |n> \left(\frac{\bar{n}^n e^{-\bar{n}}}{n!}\right)^{\frac{1}{2}} e^{-in\omega t} \tag{5.3}$$

with probability of finding n photons in the field

$$p(n) = |<\alpha|n>|^2 = \bar{n}^n e^{-\bar{n}}/n! \, . \tag{5.4}$$

The mean number of photons in the field is $\bar{n}$. The expectation value of the single-mode electric field operator in a coherent state is (using 4.12)

$$\langle\alpha|\hat{\underline{E}}|\alpha\rangle = i\left(\frac{\hbar\omega_k}{2}\right)^{\frac{1}{2}} \underline{u}_k(\underline{r}) \langle\alpha|\hat{a}_k\, e^{-i\omega_k t}|\alpha\rangle + \text{c.c.} \tag{5.5}$$

But

$$\hat{a}_k|\alpha\rangle = \sum_n \left(\frac{\bar{n}^n e^{-\bar{n}}}{n!}\right)^{\frac{1}{2}} \sqrt{n}\, |n-1\rangle$$

$$= \sqrt{\bar{n}}\, |\alpha\rangle e^{-i\phi} \tag{5.6}$$

where ϕ is an arbitrary phase. The expectation value of the electric field operator in a coherent state

$$\langle\alpha|\hat{\underline{E}}|\alpha\rangle = i\left(\frac{\hbar\omega_k}{2}\right)^{\frac{1}{2}} \underline{u}_k(\underline{r})\sqrt{\bar{n}}\, e^{-i\omega_k t - i\phi} + \text{c.c.}$$

$$\equiv \underline{A}_o\sqrt{\bar{n}}\, \sin(\omega t + \phi) \tag{5.7}$$

looks just like a classical coherent field of amplitude $A_o\sqrt{\bar{n}}$. The field amplitude is proportional to the square root of the mean number of photons. We recall that we can write the classical cycle-averaged field energy as

$$W = \frac{1}{2}\int \varepsilon_o |\underline{E}(\underline{r},t)|^2\, d^3\underline{r}$$

in terms of the mean number of photons in the mode

$$W = \bar{n}\hbar\omega \quad .$$

The coherent sinusoidally varying field described in (5.7) is just what one might have postulated from semi-classical theory by equating these two expressions for W. In this chapter we explore the properties of the coherent state and some of its uses in quantum optics.

5.2 COHERENT STATES

We saw in Chapter 4 that a field is fully coherent if the radiation field is perfectly monochromatic with well-defined amplitude possessing no stochastic or random time evolution. Quantum mechanically this implies that there is no uncertainty about the initial wave-function: in this case the density operator describes a pure state

$$\hat{\rho} = |\ \rangle\langle\ | \ . \tag{5.8}$$

Then the coherence functions are

$$G^n(x_1;\text{---}x_{2n}) = \langle|\hat{E}^{(-)}(x_1)\text{---}\hat{E}^{(-)}(x_n)\hat{E}^{(+)}(x_{n+1})\text{---}\hat{E}^{(+)}(x_{2n})|\rangle \tag{5.9}$$

If there existed a pure state in which $\hat{E}^{(-)}$ and $\hat{E}^{(+)}$ were simultaneously diagonal then $G^{(n)}(x_1;\text{--}x_{2n})$ would factorize. However this cannot happen because $\hat{E}^{(-)}$ and $\hat{E}^{(+)}$ do not commute; $[\hat{E}^{(-)}, \hat{E}^{(+)}] \neq 0$. But, if we demand that $|\ \rangle$ is a right eigenstate of $\hat{E}^{(+)}(x)$ with eigenvalue $\mathcal{E}(x)$

$$E^{(+)}(x)|\ \rangle = \mathcal{E}(x)|\ \rangle \tag{5.10}$$

together with the dual relationship

$$\langle\ |E^{(-)}(x) = \mathcal{E}^*(x)\langle\ | \tag{5.11}$$

then $G^{(n)}$ readily factorizes as required, although again we should note that normal ordering was essential. The right eigenstates of $\hat{E}^{(+)}(x)$ are the coherent states; these are quite different to the number states which are energy eigenstates. The coherent state eigenvalue equation (5.7) treats fields as observables whereas we know from Section (5.1) that $\langle|E^{(+)}|\rangle$ vanishes in a pure number state. The coherent states are superpositions of number states with an indefinite number of photons. Consequently making an observation of the field leaves the field in the same state even though a photon may be absorbed in the act of observation. This follows because $E^{(+)}$ is proportional to $\hat{a}$, the annihilation operator, and applying $\hat{a}$ to the coherent state generates precisely the same state.

We return to the normal mode expansion and ignoring the vector character of the field, write

$$\hat{E}^{(+)}(\underline{r},t) = i\sum_k \left(\frac{\hbar\omega_k}{2}\right)^{\frac{1}{2}} \hat{a}_k\, u_k(\underline{r})\, e^{-i\omega_k t} \tag{5.12}$$

so that from eq.(5.10) the eigenvalue is

$$\mathcal{E}(\underline{r},t) = i\sum \left(\frac{\hbar\omega_k}{2}\right)^{\frac{1}{2}} \alpha_k\, u_k(\underline{r})\, e^{-i\omega_k t} \tag{5.13}$$

so that

$$\hat{a}_k|\alpha_k\rangle = \alpha_k|\alpha_k\rangle\ . \tag{5.14}$$

A multimode coherent state $|\{\alpha_k\}\rangle$ is simply a direct product of single mode states, $|\alpha_k\rangle$,

$$|\{\alpha_k\}\rangle = \Pi_k|\alpha_k\rangle \tag{5.15}$$

5.3 SINGLE MODE COHERENT STATES

The normalization, orthogonality and fluctuation properties of the coherent single mode state $|\alpha\rangle$ can be deduced from an expansion in terms of number states. We have

$$\hat{a}|\alpha\rangle = \alpha|\alpha\rangle \tag{5.16}$$

and taking the matrix element with state $\langle n|$ we find

$$\langle n|\hat{a}|\alpha\rangle = (n+1)^{\frac{1}{2}}\langle n+1|\alpha\rangle = \alpha\langle n|\alpha\rangle \tag{5.17}$$

following the discussion in Chapter 2 and using eq.(2.36). This recursion relation gives us

$$<n|\alpha> = \left(\frac{\alpha^n}{(n!)^{\frac{1}{2}}}\right)<0|\alpha> . \tag{5.18}$$

These are the expansion coefficients of the coherent state $|\alpha>$ in terms of the number states

$$|\alpha> = \sum_n |n><n|\alpha> = <o|\alpha> \sum_n \frac{\alpha^n}{(n!)^{\frac{1}{2}}} |n> . \tag{5.19}$$

Consequently

$$<\alpha|\alpha> = |<0|\alpha>|^2 \sum_n \frac{|\alpha|^{2n}}{n!} = |<0|\alpha>|^2 e^{|\alpha|^2} \tag{5.20}$$

and $|\alpha>$ is normalised if we choose

$$<0|\alpha> = e^{-\frac{1}{2}|\alpha|^2} . \tag{5.21}$$

Finally we obtain the number state expansion of the coherent states

$$|\alpha> = e^{-\frac{1}{2}|\alpha|^2} \sum_n \frac{\alpha^n}{(n!)^{\frac{1}{2}}} |n> . \tag{5.22}$$

We see at once that the probability p(n) of finding n photons in state is a Poisson distribution with mean value

$$\bar{n} = |\alpha|^2 \tag{5.23}$$

with

$$p(n) = |<n|\alpha>|^2 = \frac{|\alpha|^{2n}}{n!} e^{-|\alpha|^2} . \tag{5.24}$$

The variance in the photon number, Δn, can be calculated using the Poisson distribution given by eq.(5.24). We define

$$(\Delta n)^2 \equiv \sum_n (n - \bar{n})^2 p_n = <\hat{n}^2> - <\hat{n}>^2 . \tag{5.25}$$

We write $\hat{n}^2 = \hat{a}^+\hat{a}\ \hat{a}^+\hat{a}$ in normal order by using the basic commutation relation eq.(2.16) as $\hat{a}^+(\hat{a}^+\hat{a} + 1)\hat{a} = \hat{a}^+\hat{a}^+\hat{a}\hat{a} + \hat{a}^+\hat{a}$. Then using eqs.(5.16) and (5.22) we find for the coherent state

$$<\alpha|\hat{n}^2|\alpha> = |\alpha|^4 + |\alpha|^2 \tag{5.26}$$

so that there are no "wave fluctuations" in the photon variance

$$(\Delta n)^2 = |\alpha|^2 = \bar{n} . \tag{5.27}$$

Compared with other common eigenstates the coherent states behave improperly. For example, they are not orthogonal, and the scalar product is

$$\langle\alpha|\beta\rangle = \sum_n \frac{(\alpha^*)^n\beta^n}{n!} e^{-\frac{1}{2}|\alpha|^2} e^{-\frac{1}{2}|\beta|^2}$$

$$= \exp(\alpha^*\beta - \tfrac{1}{2}(|\alpha|^2 + |\beta|^2)) \ . \tag{5.28}$$

If $(\alpha-\beta)$ is large, then the states are "almost" orthogonal because $|\langle\alpha|\beta\rangle|^2 = \exp-(|\alpha-\beta|^2)$.

The expectation value of the normally-ordered product of creation and annihilation operators is

$$\langle\alpha|(\hat{a}^+)^n \hat{a}^m|\alpha\rangle = (\alpha^*)^n\alpha^m \tag{5.29}$$

which is obtained by repeated application of eq.(5.16) and its conjugate. An alternative form of this equation is the vacuum expectation value,

$$\langle 0|(\hat{a}^+ + \alpha^*)^n(\hat{a} + \alpha)^m|0\rangle = (\alpha^*)^n\alpha^m \tag{5.30}$$

as $\hat{a}|0\rangle = 0$. This simple and obvious formula will prove to be extraordinarily useful when we come to discuss the semi-classical limit in Chapter 6. In fact the coherent state expectation of any function of the operators $\hat{a}$ and $\hat{a}^+$ can be written in terms of vacuum expectations if the operators are normally ordered

$$\langle\alpha|F(\hat{a},\hat{a}^+)|\alpha\rangle = \langle 0|F(\hat{a} + \alpha, \hat{a}^+ + \alpha^*)|0\rangle \ . \tag{5.31}$$

5.4 COHERENT STATES AS MINIMUM UNCERTAINTY STATES

Rather than concentrate on the non-Hermitian operators $\hat{a}$ and $\hat{a}^+$ unduly, we return to Hermitian position $\hat{q}$ and momentum operators $\hat{p}$ of the single-mode field effective oscillator

$$\hat{q} = (\hbar/2\omega)^{\frac{1}{2}} (\hat{a}^+ + \hat{a}) \tag{5.32}$$

$$\hat{p} = i(\frac{\hbar\omega}{2})^{\frac{1}{2}} (\hat{a}^+ - \hat{a}) \ . \tag{5.33}$$

The coherent states in fact minimise the uncertainty product of the variables p and q; it was this fact that interested Schrödinger in the early days of quantum physics when he discovered the coherent states, then termed minimum uncertainty states (see for example Sargent, Scully and Lamb 1974). The variances in the position and momentum are

$$(\Delta q)^2 = \langle\hat{q}^2\rangle - \langle\hat{q}\rangle^2 \tag{5.34}$$

$$(\Delta p)^2 = \langle\hat{p}^2\rangle - \langle\hat{p}\rangle^2 \ . \tag{5.35}$$

The expectation of $\hat{p}$ and $\hat{q}$ can be found from eqs.(5.32, 5.33) and the definition (5.16) to be,

$$\langle\alpha|\hat{q}|\alpha\rangle = \left(\frac{2\hbar}{\omega}\right) \text{Re}\ \alpha \tag{5.36}$$

$$\langle\alpha|\hat{p}|\alpha\rangle = (2\hbar\omega)\ \text{Im}\ \alpha . \tag{5.37}$$

The moments $\langle\alpha|\hat{q}^2|\alpha\rangle$ and $\langle\alpha|\hat{p}^2|\alpha\rangle$ can be computed using the equivalent of eq.(5.26) to give

$$(\Delta q)^2 = \hbar/2\omega \tag{5.38}$$

$$(\Delta p)^2 = \hbar\omega/2 \tag{5.39}$$

such that the product

$$\Delta q\ \Delta p = \hbar/2 \tag{5.40}$$

is the minimum permitted by Heisenberg's Uncertainty Principle.

A single mode field prepared initially in a coherent state evolves as a coherent state according to the time-dependent Schrödinger equation

$$i\hbar \frac{\partial}{\partial t} \psi(t) = \hat{\mathcal{H}}\psi(t) \tag{5.41}$$

which is formally integrated to give

$$\psi(t) = \exp(-i\hat{\mathcal{H}}t/\hbar)\psi(0) . \tag{5.42}$$

The initial condition is $\psi(0) = |\alpha\rangle$, which is expanded in number states in eq.(5.22). At time t, the wave-function has evolved to

$$\psi(t) = \exp(-i\hat{\mathcal{H}}t/\hbar)|\alpha\rangle \tag{5.43}$$

which, on using eq.(5.22) and the relation $\exp(-i\hat{\mathcal{H}}t/\hbar)|n\rangle = \exp(-in\omega t)|n\rangle$ can be rewritten as

$$\psi(t) = |\alpha(t)\rangle \tag{5.44}$$

where $\alpha(t) = \alpha\exp(-i\omega t)$. The initial coherent state, unperturbed by any other coupling remains coherent for all times.

The energy of a single mode coherent state is

$$\langle\alpha|\hat{\mathcal{H}}|\alpha\rangle = \hbar\omega|\alpha|^2 = \hbar\omega\bar{n} . \tag{5.45}$$

The coherent state expectation value of any operator $\hat{A}$ produces a generating function for all the number state matrix elements of $\hat{A}$:

$$\langle\alpha|\hat{A}|\alpha\rangle = \sum_{m,n} e^{-|\alpha|^2} \frac{(\alpha^*)^n\alpha^m}{(n!\ m!)^{\frac{1}{2}}} \langle n|\hat{A}|m\rangle \tag{5.46}$$

while the completeness of the coherent states is given by (Glauber 1965) as

$$\frac{1}{\pi}\int d^2\alpha|\alpha\rangle\langle\alpha| = 1 . \tag{5.47}$$

5.5 QUANTUM FIELD QUASIPROBABILITIES

In this section we develop the quantum analogue of the probability distribution $P(\{C_k\})$ for the Fourier coefficients of the field which we developed in the classical description of field fluctuations. Any state describing a single mode of the radiation field can be expanded in terms of the coherent states

$$|\phi> = \frac{1}{\pi}\int |\alpha><\alpha| d^2\alpha |\phi> \tag{5.48}$$

using eq.(5.47) for the resolution of unity. The density operator $\hat{\rho}$ can also be expressed in terms of the diagonal representation

$$\hat{\rho} = \int P(\alpha) |\alpha><\alpha| d^2\alpha \tag{5.49}$$

which is sometimes called the P representation (see for example Glauber 1965, 1970 or Sargent, Scully and Lamb 1974). The multimode generalization of eqs.(5.48) and (5.49) may be obtained using (5.13). As $Tr\hat{\rho} = 1$, we have the normalisation

$$\int P(\alpha) d^2\alpha = 1 \; . \tag{5.50}$$

For example if we imagine a single-mode radiation field prepared in a coherent state $|\beta>$,

$$\hat{\rho} = |\beta><\beta| \tag{5.51}$$

we find that

$$P(\alpha) = \delta^2(\alpha - \beta) \tag{5.52}$$

where the two-dimensional delta-function is

$$\delta^2(\alpha - \beta) = \delta(Re(\alpha - \beta))\delta(Im(\alpha - \beta)) \; . \tag{5.53}$$

Using our result for the expectation value of a normally ordered product of operators, eq.(5.29), we see that,

$$\begin{aligned} Tr(\hat{\rho}(\hat{a}^+)^m(\hat{a})^n) &= \int P(\alpha) <\alpha| (\hat{a}^+)^m (\hat{a})^n |\alpha> d^2\alpha \\ &= \int P(\alpha) (\alpha^*)^m (\alpha)^n \, d^2\alpha \end{aligned} \tag{5.54}$$

which looks remarkably like averaging a classical function over a probability distribution. However $P(\alpha)$ is not a probability density, although it behaves like one for many purposes, because α is the non-observable eigenvalue of a non-Hermitian operator $\hat{a}$. Indeed it is possible for $P(\alpha)$ to be singular or even negative. For this reason $P(\alpha)$ is called a "quasiprobability". The purely quantum aspects of coherence and of $P(\alpha)$ are discussed by Kimble, Dagenais and Mandel and by Walls (Reprinted Papers 16, 13).

The multimode generalisation of the P-representation is straightforward. The multimode eigenvalue equation is

$$E^{(+)}(\underline{r},t)|\{\alpha_k\}> = \mathcal{E}(\underline{r},t;\{\alpha_k\})|\{\alpha_k\}> \tag{5.55}$$

we can use this formal apparatus to construct expressions for the quantum coherence functions as integrals over quasiprobabilities in direct analogy with the classical coherence functions eq.(4.11).

For example, the first-order coherence function is

$$G^{(1)}(x_1;x_2) = \int P(\{\alpha_k\} \mathcal{E}^*(x_1,\{\alpha_k^*\}) \mathcal{E}(x_2,\{\alpha_k\}) \Pi_k d^2\alpha_k \ . \qquad (5.56)$$

This looks just like the classical coherence function if the electric field eigenvalue $\mathcal{E}(x,\{\alpha_k\})$ replaces the classical field $E^{(+)}(x,\{C_k\})$ in eq.(4.11) and the quasiprobability $P(\{\alpha_k\})$ replaces the classical probability $P(\{C_k\})$. Nevertheless the expression in eq.(5.56) is fully quantum electrodynamic and does not depend on any ill-defined large quantum number limit to justify its classical appearance.

An example of a state of the quantized radiation field which possesses a well-behaved P-representation is a single-mode chaotic, or thermal field. The classical description of such a field may be given in terms of a Gaussian probability distribution of electric field strengths with a random-walk phase (e.g. Loudon, 1973). The P-representation is similarly a Gaussian (see for example Sargent, Scully and Lamb, 1974)

$$P(\alpha) = \frac{1}{\pi\langle n\rangle} \exp\left(-|\alpha|^2/\langle n\rangle\right) \qquad (5.57)$$

and the chaotic-field density operator is

$$\hat{\rho} = \frac{1}{\pi\langle n\rangle} \int e^{-|\alpha|^2/\langle n\rangle} |\alpha\rangle\langle\alpha| d^2\alpha \ . \qquad (5.58)$$

Alternatively this can be expressed in a number state basis as

$$\hat{\rho} = \frac{1}{1+\langle n\rangle} \sum_m \left(\frac{\langle n\rangle}{1+\langle n\rangle}\right)^m |m\rangle\langle m| \ . \qquad (5.59)$$

The probability of finding n photons in the mode is just the Bose-Einstein factor

$$P(n) = \frac{1}{1+\langle n\rangle} \left(\frac{\langle n\rangle}{1+\langle n\rangle}\right)^n \ . \qquad (5.60)$$

The number of photons of frequency ν collected in a coherence time by a detector whose area equals the coherence area of the field has been termed the photon degeneracy, δ, by Mandel. He argues that for a polarized beam of light such photons are in the same quantum mechanical state. For thermal radiation from a conventional light source such as an incandescent lamp $\delta \approx 3.10^{-4}$. Clearly the observation time for interference experiments with such fields will be exceedingly long because the probability of detecting a photon in a particular mode is minute. In contrast laser radiation contains many photons in a coherence volume.

5.6 THE GENERATOR OF COHERENT STATES

It is possible to generate the coherent states by a unitary transformation, or displacement, of the vacuum state (Glauber 1965). This method proves useful in understanding the relationship between semi-classical and fully quantized theories of the interaction of light with atoms (Mollow 1975, Pegg 1980) and will be used for this purpose in Chapter 6.

A unitary transformation simplifies many dynamical problems by changing the reference frame. In the old reference frame, the state $|t\rangle$ obeys the Schrödinger evolution equation

$$i\hbar \frac{\partial}{\partial t} |t\rangle = \hat{\mathcal{H}} |t\rangle. \tag{5.61}$$

In the new reference frame, the new state is

$$|\bar{t}\rangle = \hat{T}(t) \quad |t\rangle \tag{5.62}$$

where T(t) is the unitary transformation. In the new frame, the Schrödinger equation is

$$i\hbar \frac{\partial}{\partial t} |\bar{t}\rangle = \bar{H} \; |\bar{t}\rangle \tag{5.63}$$

or using equation (5.62)

$$i\hbar \; \dot{\hat{T}} \; |t\rangle + i\hbar \; \hat{T} \frac{\partial}{\partial t} |t\rangle = \bar{H} \; \hat{T} \; |t\rangle \tag{5.64}$$

Then from eq.(5.61) we find

$$i\hbar\dot{\hat{T}} + \hat{T}\hat{\mathcal{H}} = \bar{\mathcal{H}}\hat{T}$$

or

$$\bar{\mathcal{H}} = \hat{T}\hat{\mathcal{H}}\hat{T}^{-1} + i\hbar\dot{\hat{T}}\hat{T}^{-1} \quad . \tag{5.65}$$

Note the appearance of the second term in eq.(5.65) for the transformed Hamiltonian which is important if the transformation $\hat{T}$ is time dependent.

A unitary form for $\hat{T}$ can be written

$$\hat{T} = \exp(\hat{A}) \tag{5.66}$$

where $\hat{A}$ is some operator. Any operator $\hat{B}$ representing an observable of the system can be transformed by the action of $\hat{T}$. If, for example, $\hat{A} = i\,\hat{p}_x\,\alpha$ where α is some c-number and $\hat{p}_x$ is the linear momentum operator in the x-direction, then the position operator $\hat{x}$ is displaced as follows

$$\bar{x} = \hat{T}\,\hat{x}\,\hat{T}^{-1} = \hat{x} + \alpha \quad . \tag{5.67}$$

This can be verified by expanding the exponential in eq.(5.66) and using the commutator $[i\hat{p}_x, \hat{x}] = \hbar$. Similarly, if $\hat{A} = i\hat{x}\alpha$, then the momentum is translated to

$$\bar{p}_x = \hat{T}\,\hat{p}_x\,\hat{T}^{-1} = \hat{p}_x - \alpha \; . \tag{5.68}$$

In this way the canonical momentum and position of an oscillator can be displaced. A combination of these transformations can be used to translate the annihilation or creation operators $\hat{a}$ and $\hat{a}^+$ of the oscillator or of a single mode field defined by

$$\hat{a} = (2\hbar\omega)^{-\frac{1}{2}}(\omega\hat{x} + i\hat{p}_x) \tag{5.69}$$

and

$$\hat{a}^+ = (2\hbar\omega)^{-\frac{1}{2}}(\omega\hat{x} - i\hat{p}_x) \tag{5.70}$$

as in Chapter 2.

The operator used by Glauber is

$$\hat{D}(\alpha) = \exp(\alpha\hat{a}^+ - \alpha^*\hat{a}) \tag{5.71}$$

which transforms the annihilation and creation operators as

$$\hat{D}^{-1}(\alpha)\ \hat{a}\ \hat{D}(\alpha) = \hat{a} + \alpha \tag{5.72}$$

$$\hat{D}^{-1}(\alpha)\ \hat{a}^+\hat{D}(\alpha) = \hat{a}^+ + \alpha^* \ . \tag{5.73}$$

This can be proved either by using eqs.(5.57) - (5.70) or by using the basic commutation relations and a power series expansion.

The unitary transformation $\hat{D}(\alpha)$ not only transforms the operators $\hat{a}$ and $\hat{a}^+$ but also the radiation field states. A single-mode coherent state $|\alpha>$ is transformed to a new state $|\bar{\alpha}>$ which we determine using eqs.(5.72) and (5.73). The original states $|\alpha>$ obey the eigenvalue equation

$$\hat{a}|\alpha> = \alpha|\alpha> \tag{5.74}$$

which can be re-written in the transformed frame using $\hat{T} = \hat{D}^{-1}(\alpha)$ as

$$\hat{D}^{-1}(\alpha)\,\hat{a}\,\hat{D}(\alpha)\ \hat{D}^{-1}(\alpha)|\alpha> = \alpha\hat{D}^{-1}(\alpha)|\alpha>$$

which gives from eq.(5.72)

$$(\hat{a} + \alpha)\ \hat{D}^{-1}(\alpha)|\alpha> = \alpha\hat{D}^{-1}(\alpha)|\alpha>$$

and so

$$\hat{a}\ \hat{D}^{-1}(\alpha)|\alpha> = 0 \ .$$

It follows that,

$$\hat{D}^{-1}(\alpha)|\alpha> = |0> \tag{5.75}$$

and the transformed coherent state is the vacuum state. The inverse

$$\hat{D}(\alpha)|0> = |\alpha> \tag{5.76}$$

demonstrates that the coherent state is generated from the vacuum state by the transformation $\hat{D}(\alpha)$.

Use of the transformation $\hat{D}(\alpha)$ demonstrates how a coherent state of the electromagnetic field can be produced. Imagine a classical external current $\underline{j}(\underline{r},t)$ which is coupled to the quantized electromagnetic field $\hat{\underline{A}}(\underline{r},t)$ through the interaction

$$\hat{V}(t) = -\int \underline{j}(\underline{r},t).\hat{\underline{A}}(\underline{r},t)d^3r \ . \tag{5.77}$$

Then in the interaction representation, the state-vector evolves as

$$i\hbar \frac{\partial}{\partial t}|t> = \hat{V}(t)|t> \tag{5.78}$$

with the state at time t, $|t>$ evolving from the initial state $|i>$ at $t = 0$. We will imagine that the initial state is the vacuum field state $|0>$. The eq.(5.78) can be integrated directly to give

$$|t> = \exp\left[\frac{i}{\hbar}\int_0^t dt'\hat{V}(t') + i\theta(t)\right]|0> \tag{5.79}$$

where $\theta(t)$ is an irrelevant phase-term. The field $\hat{\underline{A}}(\underline{r},t)$ can be expanded into normal modes

$$\hat{\underline{A}}(\underline{r},t) = \sum_k \left[\hat{a}_k \underline{f}_k(\underline{r},t) + \hat{a}_k^+ \underline{f}_k^*(\underline{r},t)\right] \tag{5.80}$$

and if we define

$$\alpha_k(t) = \frac{i}{\hbar}\int_0^t dt' \int d^3r \ \underline{j}(\underline{r},t') \ \underline{f}_k^*(\underline{r},t') \tag{5.81}$$

we have

$$|t> = \prod_k \exp(\alpha_k \hat{a}_k^+ - \alpha_k^* \hat{a}_k)|0> \equiv |\{\alpha_k\}> \ . \tag{5.82}$$

We see that the field radiated by a classical current is a multimode coherent state $|\{\alpha_k\}>$. To some extent a laser operates as a macroscopic oscillating antenna with a well-defined frequency analogous to a single frequency classical current $\underline{j}(\underline{r},t)$. In this way the laser excites a single mode field which approximates to a coherent state $|\alpha>$.

5.7 HANBURY-BROWN AND TWISS INTENSITY CORRELATION

The conventional first-order coherence experiments, which are a measure of the degree of monochromaticity or of "coherence length" of the light, cannot distinguish between states of the same spectral distribution with quite dissimilar statistical properties. An example of such states would be coherent light from a laser above threshold and thermal chaotic light which is passed through a filter to give the same spectrum. In the 1950's, Hanbury-Brown and Twiss (see Reprinted Paper 14) developed a new kind of correlation experiment for applications in intensity-correlation stellar interferometry (see Hanbury-Brown 1974 for a full discussion). The principle of the experiment

is shown in Fig. (5.1)

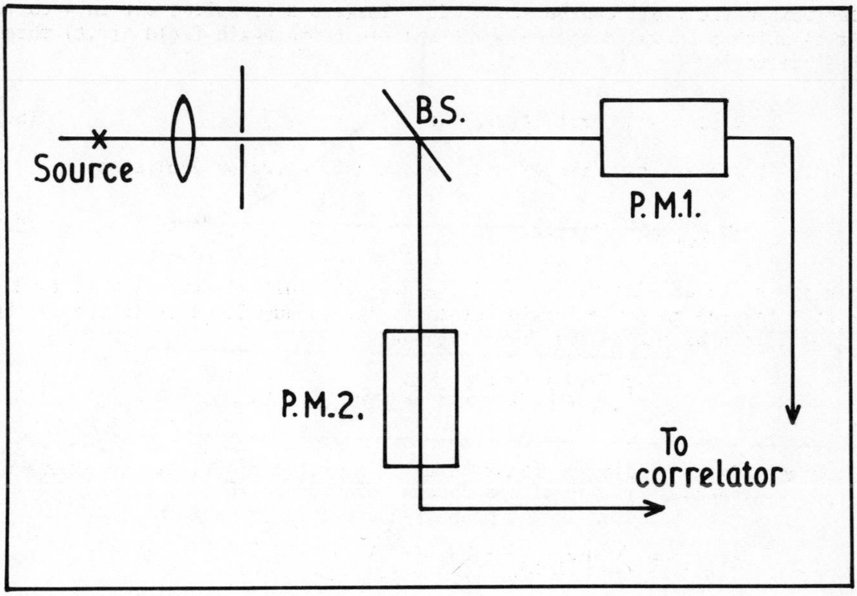

Fig. (5.1) Schematic diagram of the Hanbury-Brown Twiss intensity correlation experiment.

Narrow bandwidth light is needed for the experiment to be feasible (see Reprinted Paper 15 by Purcell). This light is collimated by a pinhole and is beam-split by a half-silvered mirror, BS. Some light is transmitted to the photomultiplier detector PM1 and some is reflected to the photomultiplier detector PM2. The coincidence rate in the detection response is then plotted against a time delay τ.

If the light incident on one photomultiplier was statistically independent of the other, a uniform coincidence rate independent of τ would be obtained. When a thermal light source, such as a discharge lamp, is used a coincidence rate at zero delay τ is produced which is twice as large as the steady rate at large τ. It is as if the photons arrived "bunched" in pairs (Fig. 5.2). This intensity correlation is called the Hanbury-Brown Twiss effect.

A well-stabilized laser source, run well above threshold, would give no such correlation. For such a laser field the Glauber coherence function $G^{(2)}$ factorizes

$$G^{(2)}(x_1;x_1;x_1;x_1) = G^{(1)}(x_1;x_1)\, G^{(1)}(x_1;x_1)$$

reflecting the fact that such a field is fully coherent in the sense of eq. (4.43).

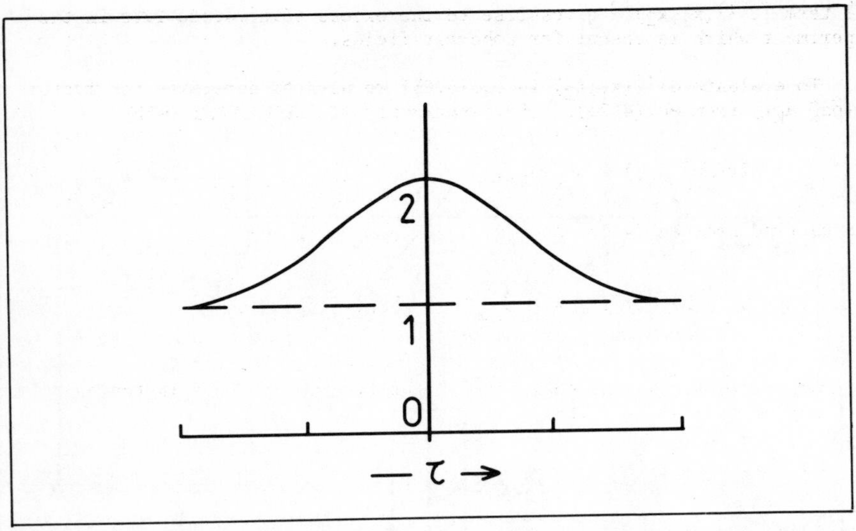

Fig. (5.2) Normalized photon coincidence rate as a function of delay in the Hanbury-Brown Twiss experiment.

A semi-classical interpretation of the experiment is straightforward (see Mandel and Wolf 1965) and rather simpler than an interpretation involving photons. We show here that the diagonal coherent state P-representation allows us to analyse the Hanbury-Brown Twiss experiment quantum mechanically using notions very close to those used in the classical approach. A multimode chaotic or thermal field has a P-representation from (5.57)

$$P(\{\alpha_k\}) = \prod_k \frac{1}{\pi\langle n_k\rangle} e^{-|\alpha_k|^2/\langle n_k\rangle} . \tag{5.83}$$

If we know the average photon number in each mode, or equivalently have knowledge of the spectrum, $\langle n_k\rangle$, we can construct $P(\{\alpha_k\})$ and from this Gaussian probability deduce all the Glauber coherence functions (e.g. eq.5.56). In this way we can see that the spectrum of chaotic light given by $\langle n_k\rangle$ as a function of mode frequency contains all the information necessary to deduce the statistical properties of the field. This is a property of Gaussian stochastic fluctuating fields only. For such a Gaussian probability all higher-order moments can be constructed from the first-order correlation function as (Glauber 1970, p.109)

$$G^{(n)}(x_1;x_2;---x_n;y_1;y_2;---y_n) = \sum_p \prod_{j=1}^{n} G^{(1)}(x_j;y_{p_j}) . \tag{5.84}$$

The joint counting rate appropriate to the Hanbury-Brown Twiss effect is from eq.(5.84)

$$G^{(2)}(x_1x_2;x_2x_1) = G^{(1)}(x_1;x_1)G^{(1)}(x_2;x_2) + |G^{(1)}(x_1;x_2)|^2 \tag{5.85}$$

The term $|G^{(1)}(x_1;x_2)|^2$ gives rise to the excess coincidence rate in the experiment which is absent for coherent fields.

To evaluate $G^{(1)}(x_1;x_2)$ in eq.(5.85) we need to determine the factor $\mathrm{Tr}(\hat{\rho}\hat{a}_k^+ \hat{a}_{k'})$ from eq.(4.28). For stationary fields (Peřina 1971)

$$\mathrm{Tr}(\hat{\rho}\hat{a}_k^+ \hat{a}_{k'}) = \langle n_k\rangle \delta_{kk'} \tag{5.86}$$

and we find from eq.(4.12) and eq.(4.32) and eq.(5.86)

$$G^{(1)}(r_1t_1;r_2t_2) = \frac{1}{2}\sum_k \hbar\omega_k \langle n_k\rangle u_k^*(\underline{r}_1)u_k(\underline{r}_2)e^{i\omega_k(t_1-t_2)} . \tag{5.87}$$

We can represent the collimated light from the pinhole in the experiment as a set of plane waves with

$$u_k(\underline{r}) \simeq e^{ikx} . \tag{5.88}$$

The first order coherence function for such light is

$$G^{(1)}(\underline{r}_1t_1;\underline{r}_2t_2) = G^{(1)}(\tau) \tag{5.89}$$

where

$$\tau \equiv t_1 - t_2 - \frac{1}{c}(x_1-x_2) . \tag{5.90}$$

The emission from a Doppler-broadened spectral light source, such as that from a thermal lamp, has a Gaussian spectrum

$$\langle n_{k,\omega}\rangle = \mathrm{Const}\times\exp[-(\omega-\omega_0)^2/2\gamma^2] \tag{5.91}$$

where γ is the spectral width of the source centred at frequency ω_0, so that

$$G^{(1)}(\tau) = G^{(1)}(0)\exp(i\omega_0\tau - \tfrac{1}{2}\gamma^2\tau^2) . \tag{5.92}$$

The second-order, or intensity, correlation function is then

$$\begin{aligned} G^{(2)}(x_1;x_2;x_2;x_1) &= G^{(1)}(x_1;x_1)G^{(1)}(x_2;x_2) + |G^{(1)}(x_1;x_2)|^2 \\ &= G^{(1)}(0)G^{(1)}(0) + |G^{(1)}(0)|^2\exp(-\gamma^2\tau^2) \\ &= |G^{(1)}(0)|^2(1+\exp(-\gamma^2\tau^2)) \end{aligned} \tag{5.93}$$

exactly as shown in Fig. (5.2). The Hanbury-Brown Twiss effect shows that the normalized correlation function which is a measure of the coincidence rate, is not small but increases from 1 at large τ to 2 at $\tau = 0$, with a large correlation for a short time $\sim\gamma^{-1}$. This is why narrow-band radiation is necessary for the observation of the effect. In practice, detector efficiencies, rise times and resolving times reduce the size of the coincidence rate.

Glauber (1970) has given a helpful heuristic explanation of the effect. Imagine the light to be a single mode field; the mode distribution tells us how fast the correlation falls off with τ, not the size of the peak of the correlation. Then for zero time interval, using (5.56),

$$G^{(1)}(0) = \int P(\alpha)|\mathcal{E}|^2 d^2\alpha$$
$$= \text{const} \int \exp(-|\alpha|^2/\langle n\rangle)|\alpha|^2 d^2\alpha \quad . \tag{5.94}$$

The function $G^{(2)}$ is the fourth moment of the field:

$$G^{(2)}(0) = \int P(\alpha)|\mathcal{E}|^4 d^2\alpha = 2\{G^{(1)}(0)\}^2 \tag{5.95}$$

using eq. (5.93). Gaussian functions have the property that the fourth moment is twice the square of the second moment. Because the Gaussian has a long tail, the higher-order correlation functions evaluated for zero time arguments are larger by a factor of n! than the nth power of the first-order correlation function. For a coherent field, eq. (5.52) gives us a delta-function with no tail at all. One could imagine a field from a well-stabilized laser having a distribution which approaches this form (Fig. 5.3), and

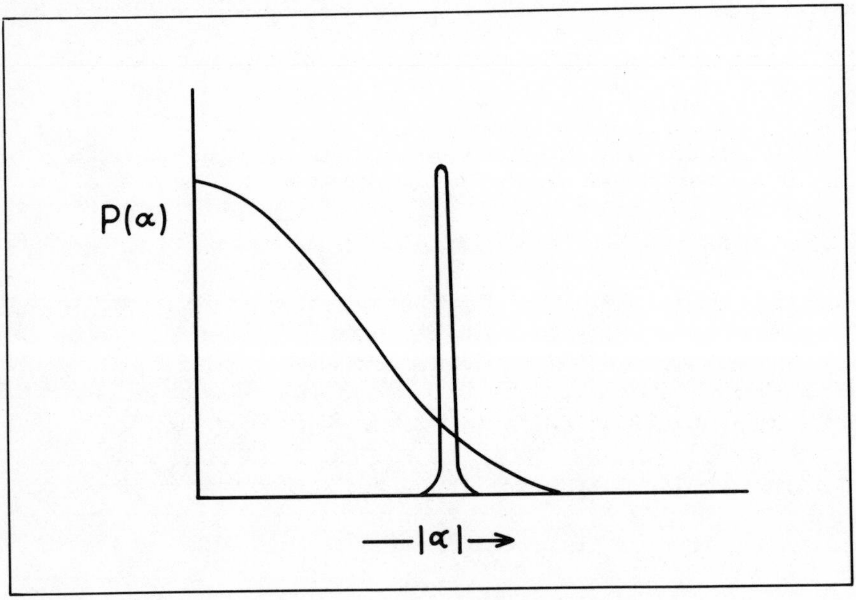

Fig. 5.3 Comparison of the diagonal coherent state P-representation for chaotic and coherent light as a function of $|\alpha|$

for such a distribution the fourth moment is simply the square of the second moment and gives no enhancement effect.

In a chaotic field the light amplitude fluctuates rapidly. A pair of photon counters in such a field will both tend to register when the amplitude is large, but neither will tend to register if the amplitude is small. This leads to the coincidence rate, since the counters tend to register together and tend to keep quiet together. No such correlation between counters is produced by a coherent field because the counters see the same amplitude, constant in time, and their behaviour is statistically uncorrelated. It is very hard to explain the correlation using wave-packet, or "billiard-ball" notions of photons, and attempts to do so led to great confusion when the Hanbury-Brown Twiss effect was first discovered.

CHAPTER 6

Semi-classical Theory and Quantum Electrodynamics

6.1 SEMI-CLASSICAL LIMITS

The semi-classical theory of the interaction of quantized matter with classical radiation fields is simple but applicable to much of quantum optics. In this chapter we survey the need for field quantization and discuss the breakdown of semi-classical theories. For further reading we recommend the reviews by Milonni (1976), Mandel (1976) and Loudon (1978). The semi-classical approximation is not a large quantum number approximation as it successfully describes the interference effects produced by thermal light sources for which the occupation number n is small with $n \sim 10^{-3}$. We will demonstrate the connection between the breakdown of semi-classical theory and the role of spontaneous emission. More generally, the semi-classical theory breaks down for all effects which depend intimately on the field commutator $[\hat{a},\hat{a}^{+}] = 1$. The existence of a well-behaved diagonal coherent state representation of the field density matrix discussed in Chapter 5 is connected with this commutator but this is a technical point which is really beyond the scope of this text.

6.2 THE RELATION BETWEEN SEMI-CLASSICAL AND QUANTIZED HAMILTONIANS

It is possible to relate the semi-classical and fully-quantized Hamiltonians describing the interaction of atoms with radiation fields quantitatively using the Glauber displacement operator $\hat{D}(\alpha)$ (Mollow 1975). Here we follow the discussion of Pegg (1980). The semi-classical Hamiltonian is

$$\hat{\mathcal{H}}_{SC} = \hat{\mathcal{H}}_a - \underline{\hat{d}}.\underline{E}(t) \tag{6.1}$$

where $\hat{\mathcal{H}}_a$ is the free atom Hamiltonian and $\underline{d}$ the electric dipole operator. The fully quantized Hamiltonian describing the coupling of the atom to a single mode field in the interaction representation, where the field operators have their unperturbed time-evolution, is

$$\hat{\mathcal{H}}_q = \hat{\mathcal{H}}_a - \underline{\hat{d}}.\underline{\hat{E}}(t) \; , \tag{6.2}$$

where

$$\underline{\hat{E}}(t) = i(\hbar\omega/2\varepsilon_0 V)^{\frac{1}{2}}\underline{\varepsilon}_k[\hat{a}e^{-i\omega t} - \hat{a}^{+}e^{i\omega t}] \; . \tag{6.3}$$

The more usual Schrödinger picture Hamiltonian is obtained from eq.(6.2) by a unitary transformation as in Chapter 4 with a transformed state vector

$$|\bar{t}\rangle = \hat{T}|t\rangle \tag{6.4}$$

where

$$i\hbar \frac{\partial}{\partial t} |\bar{t}> = \bar{\mathcal{H}}|\bar{t}> . \tag{6.5}$$

The transformed Hamiltonian is

$$\bar{\mathcal{H}} = \hat{T}\hat{\mathcal{H}}\hat{T}^{-1} + i\dot{\hat{T}}\hat{T}^{-1} . \tag{6.6}$$

The specific transformation to the Schrödinger picture is obtained by using the operator

$$\hat{T} = \exp(-i\hat{\mathcal{H}}_R t/\hbar) \tag{6.7}$$

where

$$\hat{\mathcal{H}}_R = \hbar\omega(\hat{a}^+\hat{a} + \frac{1}{2}) . \tag{6.8}$$

Then $i\dot{\hat{T}}\hat{T}^{-1} = \hat{\mathcal{H}}_R$, whereas the transformation $\hat{T}\hat{\mathcal{H}}\hat{T}^{-1}$ removes the time-dependence from $\hat{E}(t)$. This can be shown by using

$$\hat{a}\exp(i\omega\hat{a}^+\hat{a}) = \exp(i\omega[\hat{a}^+\hat{a} + 1)t)\hat{a} \tag{6.9}$$

$$\hat{a}^+\exp(i\omega\hat{a}^+\hat{a}) = \exp(i\omega(\hat{a}^+\hat{a} - 1)t)\hat{a}^+ . \tag{6.10}$$

Equations (6.9) and (6.10) can be verified by series expansion. The transformed quantized Hamiltonian in the Schrödinger picture is therefore

$$\hat{\mathcal{H}}_q' = \hat{\mathcal{H}}_a - \underline{\hat{d}}.\underline{\hat{E}} + \hat{\mathcal{H}}_R . \tag{6.11}$$

We have previously described the light from a stabilized single-mode laser operating well above threshold by a coherent state $|\alpha>$. In semi-classical theory such light is represented by a monochromatic electric field. The fully quantized interaction picture Hamiltonian $\mathcal{H}_q$ can be transformed into a new picture, called the vacuum picture by Pegg, in which the nature of the semi-classical approximation is transparent. For the transformation $\hat{T}^{-1}$ in eqs.(6.4 - 6.6) we use the Glauber displacement operator (eq.5.71)

$$\hat{D}(\alpha) = \exp(\alpha\hat{a}^+ - \alpha^*\hat{a}) . \tag{6.12}$$

As we saw in Chapter 5, this transforms the annihilation and creation operators as

$$\hat{D}^{-1}(\alpha)\hat{a}\,\hat{D}(\alpha) = \hat{a} + \alpha \tag{6.13}$$

$$\hat{D}^{-1}(\alpha)\hat{a}^+\,\hat{D}(\alpha) = \hat{a}^+ + \alpha^* . \tag{6.14}$$

Using $\hat{a}|\alpha> = \alpha|\alpha>$ we find

$$\hat{D}^{-1}(\alpha)\hat{a}\,\hat{D}(\alpha)\,\hat{D}^{-1}(\alpha)|\alpha> = \alpha\hat{D}^{-1}(\alpha)|\alpha>$$

or

$$(\hat{a} + \alpha)\,\hat{D}^{-1}(\alpha)|\alpha> = \alpha\,\hat{D}^{-1}(\alpha)|\alpha>$$

so that applying the annihilation operator $\hat{a}$ to the transformed state

$\hat{D}^{-1}(\alpha)|\alpha\rangle$ gives zero. As the annihilation operator acts to reduce the occupation number by one, the transformed state must necessarily be the vacuum state $|0\rangle$ as in Section (5.6),

$$\hat{D}^{-1}(\alpha)|\alpha\rangle = |0\rangle \ . \tag{6.15}$$

The transformation $\hat{D}(\alpha)$ is time-independent so the transformed Hamiltonian in the vacuum picture is

$$\hat{\mathcal{H}}''_q = \hat{D}^{-1}(\alpha)\,\hat{\mathcal{H}}_q\,\hat{D}(\alpha) \ . \tag{6.16}$$

The new state vector in the transformed vacuum picture is $|t''_q\rangle = \hat{T}|t_q\rangle = \hat{D}^{-1}(\alpha)|\alpha\rangle = |0\rangle$. The new Hamiltonian, from eq.(6.16), is

$$\begin{aligned}\hat{\mathcal{H}}''_q &= \hat{\mathcal{H}}_a - \underline{\hat{d}}.\hat{D}^{-1}(\alpha)\underline{\hat{E}}(t)\,\hat{D}(\alpha) \\ &= \hat{\mathcal{H}}_a - \underline{\hat{d}}.\underline{\hat{E}}'(t)\end{aligned} \tag{6.17}$$

where the transformed electric field operator $\underline{\hat{E}}'(t)$ is obtained from eq.(6.3) by replacing $\hat{a}$ by $\hat{a} + \alpha$ and $\hat{a}^+$ by $\hat{a}^+ + \alpha^*$. The transformation $\hat{D}^{-1}(\alpha)$ merely adds a c-number term $\underline{E}_c(t)$ to $\underline{\hat{E}}(t)$ where

$$\underline{E}_c(t) = i(\hbar\omega/2\varepsilon_0 V)^{\frac{1}{2}}\underline{\varepsilon}(\alpha e^{-i\omega t} - \alpha^* e^{i\omega t}) \ . \tag{6.18}$$

This new classical field is the eigenvalue of the electric field operator for the state $|\alpha\rangle$, as in eq.(5.10). Therefore in the vacuum picture, an incident field in a coherent state interacting with an atom is described as the interaction of the same atom with the vacuum by a Hamiltonian

$$\hat{\mathcal{H}}''_q = \hat{\mathcal{H}}_a - \underline{\hat{d}}.(\underline{\hat{E}}(t) + \underline{E}_c(t)) \ . \tag{6.19}$$

In this picture, the incident field is modelled without approximation as a classical monochromatic field provided the atom is still coupled to the vacuum by the quantum interaction $-\underline{\hat{d}}.\underline{\hat{E}}(t)$.

The multimode generalization of this transformation is straightforward. Each occupied mode described by a coherent state can be transformed using $\hat{D}^{-1}(\alpha)$: empty modes are untransformed. More generally any field whose density matrix can be described by a diagonal coherent state representation can be transformed to the vacuum picture; all quantum effects of the radiation field are ascribed to the coupling of the atom to the vacuum field. In other words, there is no difference between an incident coherent field and a classical field plus the vacuum. The electric field strength uncertainty

$$\Delta E = \sqrt{\langle\alpha|\underline{\hat{E}}^2|\alpha\rangle - \langle\alpha|\underline{\hat{E}}|\alpha\rangle^2} = (\hbar\omega/2\varepsilon_0 V)^{\frac{1}{2}}$$

is precisely the same in both cases, see Chapter 5, so that a coherent quantum field can be regarded as a classical field with added vacuum fluctuations.

6.3 IMPORTANCE OF QUANTUM FLUCTUATIONS

The vacuum interaction is an essential ingredient in spontaneous emission. As we assumed in Chapter 4, a detector atom initially in its ground state which is only weakly coupled to the radiation field has only a small probability of excitation. The vacuum interaction is of no importance unless the atom is excited. The quantum fluctuations associated with the vacuum interaction are of no importance for weakly excited atoms. The decay from an initially excited atom, or one which is strongly driven, is triggered by the vacuum fluctuations rather as a pencil balanced vertically on its point, is triggered to fall by environmental vibrations. In consequence, the field spontaneously emitted by a decaying atom, or emitted by an atom strongly driven by an incident coherent field has pronounced quantum properties (see for example Loudon 1980, Knight and Milonni 1980, and the Reprinted Papers 12 and 13 by Loudon and Walls). Such quantum effects have been clearly observed in the experiments of Schuda, Stroud and Hercher (Reprinted Paper 11) and Kimble, Dagenais and Mandel (Reprinted Paper 16). These inherently quantum effects are quite outside of the scope of semi-classical theories and are under intense current investigation in laboratories throughout the world.

6.4 QUANTUM ASPECTS OF LIGHT

In Chapters 4 and 5 we showed how a thermal light field gave rise to photon bunching in second-order coherence experiments. In such cases $g^{(2)}(0) = 2$. A light field is said to be "antibunched" (see for example Loudon 1980, Knight and Milonni 1980) if its degree of second-order coherence is less than unity,

$$1 > g^{(2)}(\underline{r}_1 t_1;\underline{r}_1 t_1;\underline{r}_1 t_1;\underline{r}_1 t_1) \geq 0 \ . \tag{6.20}$$

The physical significance of eq.(6.20) is that the probability of detecting a coincident pair of photons from an antibunched field is less than that from a fully coherent light field with a random Poisson distribution of photons. It is as if the photon detection events had some kind of mutual repulsion. A field describable as a number state is a simple, though unrealisable, example of antibunched light with, from eq.(4.47),

$$g^{(2)}(0) = (n-1)/n \tag{6.21}$$

which is always less than unity and significantly so for small photon numbers.

Suppose an incident field described by a number state is directed at a half-silvered mirror (Loudon 1980) and a Hanbury-Brown Twiss correlation experiment performed to measure the second order coherence of the light. The normalized degree of second-order coherence $g^{(2)}(\underline{r_1}t_1;\underline{r_2}t_2;\underline{r_2}t_2;\underline{r_1}t_1)$ is given by eq.(4.45) in terms of expectation values of combinations of field mode operators

$$\hat{\underline{E}}^{(\pm)}(\underline{r},t) = i\sum_k \left(\frac{\hbar\omega_k}{2}\right)^{\frac{1}{2}} \hat{a}_k\, u_k(\underline{r})e^{-i\omega_k t} \ . \tag{6.22}$$

In our idealized Hanbury-Brown Twiss experiment we measure the field at points r_1t_1 and r_2t_2 and consider the two beams emerging from the mirror as two modes of the field which we label 1 and 2. Then the geometrical factors and constants in eq.(4.45) cancel to leave

$$g_{12}^{(2)} = \frac{<\hat{a}_1^+ \hat{a}_2^+ \hat{a}_2 \hat{a}_1>}{\bar{n}_1 \bar{n}_2} \tag{6.23}$$

where $\bar{n}_1 = <\hat{a}_1^+\hat{a}_1>$ and $\bar{n}_2 = <\hat{a}_2^+\hat{a}_2>$ are the mean photon numbers in the two beams and we write the correlation between the beams as $g_{12}^{(2)}$ for brevity. Each photon either passes through the half-silvered mirror or, with equal probability, is reflected so the mean photon count rate in each beam is $\bar{n}_i = n/2$, $i = 1,2$. Reprinted Paper 17 by Walls discusses the quantum nature of the two modes 1 and 2. The incident field is a number-state field $|n>$ described by a creation operator $\hat{a}^+$. The incident photons have equal probability of transmission or reflection at the mirror, so we can write

$$\hat{a}^+ = 2^{-\frac{1}{2}} (\hat{a}_1^+ + \hat{a}_2^+) \tag{6.24}$$

where $\hat{a}_1^+$ and $\hat{a}_2^+$ create photons in the modes 1 and 2. We know from eq.(2.37) that $|n> = (n!)^{-\frac{1}{2}}(a^+)^n|0>$, so from eq.(6.24)

$$|n> = (n!)^{-\frac{1}{2}} (2^n)^{-\frac{1}{2}} (\hat{a}_1^+ + \hat{a}_2^+)^n |0> . \tag{6.25}$$

The probability that n_1 photons are detected in beam 1 and n_2 photons are detected in beam 2, with $n_1 + n_2 = n$, is

$$P(n_1 n_2) = |<n_1,n_2|n>|^2 . \tag{6.26}$$

We use eq.(6.25) for $|n>$ and pick out the term in the expansion of $(\hat{a}_1^+ + \hat{a}_2^+)^n$ which creates n_1 photons in mode 1 and n_2 photons in mode 2, which is $(\hat{a}_1^+)^{n_1} (\hat{a}_2^+)^{n_2}$. We find as expected

$$P(n_1 n_2) = \frac{n!}{n_1! n_2! 2^n} . \tag{6.27}$$

This probability distribution can be used to obtain the correlation in photon counts in the two beams

$$<n_1 n_2> = n(n-1)/4 \tag{6.28}$$

and consequently

$$g_{12}^{(2)}(0) = (n-1)/n \tag{6.29}$$

as in the single-beam expression eq.(6.21). For example, if $n = 1$, this incident photon is either reflected or transmitted by the mirror so that n_1 can be one or zero and n_2 can be zero or one with average values $\bar{n}_1 = 1/2$, $\bar{n}_2 = 1/2$ but with $<n_1 n_2>$ equal to zero. This has been confirmed experimentally by Clauser (Reprinted Paper 8). A classical wavepacket would divide into two identical halves at the mirror so the classical degree of second-order coherence in this experiment would equal unity. The Clauser experiment is a direct confirmation of the quantum nature of light with the photon viewed as a mode occupation obeying quantum laws of probability. Reprinted Paper 12 by Loudon examines further the Hanbury-Brown Twiss experiment from a photon viewpoint.

Antibunched light can be generated experimentally from the radiation scatterd by a single atom, in a resonance fluorescence experiment, (Carmichael and Walls (1976); Kimble, Dagenais and Mandel, Reprinted Paper 16). The radiation emitted from the resonantly-excited atom is collected, split into two and imaged on to two photomultiplier tubes whose output is electronically correlated. The intensity correlation is

$$g^{(2)}(\tau) = \langle:\hat{I}(t+\tau)\hat{I}(t):\rangle/\langle\hat{I}(t+\tau)\rangle\langle\hat{I}(t)\rangle \tag{6.30}$$

where the colons denote normal ordering. The correlation at zero time delay exhibits antibunching, or negative photoelectric correlation, with

$$g^{(2)}(0) - 1 < 0 \ . \tag{6.31}$$

It is hard to imagine a well-behaved classical probability function which will lead to the mean square intensity being less than the square of the mean intensity, $\langle I^2\rangle < \langle I\rangle^2$ implied by eq.(6.31). This behaviour is strictly quantum mechanical in origin. The atom emits a photon at time $t = 0$ and is unable to re-radiate immediately after it has made a quantum jump down to the lower state. The measurement of $g^{(2)}$ consists of two parts: detection of the first photon, which ensures the de-excitation of the source atom, and a delay while the atom is re-excited by the resonant driving field for the subsequent emission of a second photon (Reprinted Paper 13 by Walls). If there is more than one atom in the observation region the interpretation is a little harder since the fluorescence could be produced by more than one atom and the antibunching would be obscured.

It is remarkable that it has been necessary to pursue such esoteric effects as strong-field resonance fluorescence to find direct evidence for the quantum nature of light. The semi-classical theory of the interaction of light with atoms is so successful and so simple that it is a temptation to use it universally in quantum optics. The quantum theory of light predicts all of these effects without much more difficulty but has an extra richness and strangeness which is only now being revealed by current research in quantum optics.

References

Allen L and Eberly J H 1975 *Optical Resonance and Two-Level Atoms*, Wiley, New York

Bethe H 1947 *Phys. Rev. 72* 339

Born M, Heisenberg W and Jordan P 1926 *Z. Phys. 35* 557

Born M and Wolf E 1964 *Principles of Optics*, 2nd Ed: Pergamon Press, Oxford

Bose S N 1924 *Z. Phys. 26* 178

Carmichael H J and Walls D F 1976 *J. Phys. B9* L43

Casimir H B G 1948 *Proc. Ned. aka Wetenschappen (Amst) 60* 793

Clauser J F 1974 *Phys. Rev. D9* 853

Dirac P A M 1927 *Proc. Roy. Soc. A114* 243

Durrant A V 1976 *Am. J. Phys. 44* 630

Einstein A 1909 *Phys. Z. 10* 185

Einstein A 1917 *Phys. Z. 18* 121

Einstein A 1924 *Berliner Berichte 1924* 261

Einstein A 1925 *Berliner Berichte 1925* 3

Feynman R P, Leighton R B and Sands M 1965 *The Feynman Lectures on Physics*. Addison Wesley. Reading. See Chapter 37, Vol.I and Chapter 3, Vol.II

Fowles G R 1975 *Introduction to Modern Optics*. Holt, Rinehart and Winston. New York

Frisch O R 1965 *Contemp. Phys. 7* 45

Genesis in *The Bible* 1611 King James I Authorized Version

Glauber R 1965 in *Quantum Optics and Electronics. Les Houches 1964.* Ed: C. de Witt, A Blandin, C Cohen-Tannoudji. Gordon and Breach. New York.

Glauber R 1979 in *Quantum Optics*. Ed: S Kay and A Maitland. Academic Press. London

Haken H 1969 *Handbuch der Physik*. Vol. 25/2C. Springer-Verlag. Berlin

Haken H 1981 *Light*. Vol.I. North-Holland. Amsterdam

Hanbury-Brown R 1974 *The Intensity Interferometer*. Taylor and Francis. London

Hanbury-Brown R and Twiss R Q 1956 *Nature 177* 27

Hecht E and Zajac A 1974 *Optics*. Addison-Wesley. Reading

Heitler W 1954 *The Quantum Theory of Radiation*. 3rd Edition. O.U.P.

Jammer M 1966 *The Conceptual Development of Quantum Mechanics*. McGraw-Hill. New York

Jaynes E T 1978 in *Coherence and Quantum Optics. IV*. Ed: L Mandel and E Wolf. Plenum Press. New York

Kimble H J, Dagenais M and Mandel L 1977 *Phys. Rev. Letts. 39* 691

Knight P L and Milonni P 1980 *Phys. Rep. C. 66* 21

Lamb W E Jr. 1973 in *Impact of Basic Research on Technology*. Ed: B Kursunoglu and A Perlmutter. Plenum. New York. pp.59-111

Lamb W E Jr. and Retherford R C 1947 *Phys. Rev. 72* 241

Loudon R 1973 *The Quantum Theory of Light*. O.U.P.

Loudon R 1976 *Phys. Bull. 27* 21

Loudon R 1980 *Rep. Prog. Phys. 43* 913

Magyar G and Mandel L 1963 *Nature 198* 255

Mandel L 1961 *J.O.S.A. 51* 797

Mandel L 1976 in *Prog. in Opt. XIII*. Ed: E Wolf. p.27. North-Holland. Amsterdam

Mandel L, Sudarshan E C G and Wolf E 1964 *Proc. Phys. Soc. 84* 435

Mandel L and Wolf E 1965 *Rev. Mod. Phys. 37* 231

Matthews P T 1974 *Introduction to Quantum Mechanics*. 2nd Ed: McGraw-Hill. New York

Maxwell J C 1865 *Phil. Trans. Roy. Soc. 155* 459

Merzbacher E 1961 *Quantum Mechanics*. 2nd Ed: Wiley. New York

Milonni P W 1976 *Phys. Rep. C 25* 1

Milonni P W 1982 *Proc. Perugia Symposium on Wave-Particle Dualism*

Mollow B R 1975 *Phys. Rev. A12* 1999

Newton G, Andrews D A and Unsworth P J 1979 *Phil. Trans. Roy. Soc. 290* 373

Nussenzweig H M 1973 *Introduction to Quantum Optics*. Gordon and Breach. New York

Pais A 1979 *Rev. Mod. Phys. 21* 863

Pegg D T 1980 in *Laser Physics*. Ed: D F Walls and J D Harvey. Academic Press. Sydney

Perina J 1971 *Coherence of Light*. Van Nostrand Reinhold. New York

Pfleegor R L and Mandel L 1967 *Phys. Letts. 24A* 766

Power E A 1964 *Introductory Quantum Electrodynamics*. Longman. London

Power E A 1966 *Am. J. Phys. 34* 516

Power E A and Zienau S 1959 *Phil. Trans. Roy. Soc A251* 427

Purcell E M 1956 *Nature 178* 1449

Sargent M, Scully M O and Lamb W E Jr. 1974 *Laser Physics*. Addison-Wesley. Reading

Schiff L I 1955 *Quantum Mechanics*. McGraw-Hill. New York

Schuda F, Stroud C R Jr. and Hercher M 1974 *J. Phys. B 7* L198

Taylor G I 1909 *Proc. Camb. Phil. Soc. 15* 114

ter Haar D 1961 *Rep. Prog. Phys. 24* 304

ter Haar D 1967 *The Old Quantum Theory*. Pergamon Press. Oxford

Walls D F 1977 *Am. J. Phys. 45* 952

Walls D F 1979 *Nature 280* 451

Welton T A 1948 *Phys. Rev. 74* 1157

Whittaker E 1951 *A History of the Theories of the Aether and Electricity: The Classical Theories*. Nelson. London

Index

Reprinted Papers

Reprinted Papers

Reprinted from *Proc. Camb. Philos. Soc.*, **15** (1909), 114–115

Interference Fringes with Feeble Light

G. I. TAYLOR
Trinity College, Cambridge, UK

The phenomena of ionisation by light and by Röntgen rays have led to a theory according to which energy is distributed unevenly over the wave-front (J.J. Thomson, *Proc. Camb. Phil. Soc.* XIV. p.417, 1907). There are regions of maximum energy widely separated by large undisturbed areas. When the intensity of light is reduced these regions become more widely separated, but the amount of energy in any one of them does not change, that is, they are indivisible units.

So far all the evidence brought forward in support of the theory has been of an indirect nature; for all ordinary optical phenomena are average effects, and are therefore incapable of differentiating between the usual electromagnetic theory and the modification of it that we are considering. Sir J.J. Thomson however suggested that if the intensity of light in a diffraction pattern were so greatly reduced that only a few of these indivisible units of energy should occur on a Huygens zone at once the ordinary phenomena of diffraction would be modified. Photographs were taken of the shadow of a needle, the source of light being a narrow slit placed in front of a gas flame. The intensity of the light was reduced by means of smoked glass screens.

Before making any exposures it was necessary to find out what proportion of the light was cut off by these screens. A plate was exposed to direct gas light for a certain time. The gas flame was then shaded by the various screens that were to be used, and other plates of the same kind were exposed till they came out as black as the first plate on being completely developed. The times of exposure necessary to produce this result were taken as inversely proportional to the intensities. Experiments made to test the truth of this assumption shewed it to be true if the light was not very feeble.

Five diffraction photographs were then taken, the first with direct light and the others with the various screens inserted between the gas flame and the slit. The time of exposure for the first photograph was obtained by trial, a certain standard of blackness being attained by the plate when fully developed. The remaining times of exposure were taken from the first in the inverse ratio of the corresponding intensities. The longest time was 2000 hours or about 3 months. In no case was there any diminution in the sharpness of the pattern although the plates did not all reach the standard blackness of the first photograph.

In order to get some idea of the energy of the light falling on the plates in these experiments a plate of the same kind was exposed at a distance of two metres from a standard candle till complete development brought it up to the standard of blackness. Ten seconds sufficed for this. A simple calculation will shew that the amount of energy falling on the plate during the longest exposure was the same as that due to a standard candle burning at a distance slightly exceeding a mile. Taking the value given by Drude for the energy in the visible part of the spectrum of a standard candle, the amount of energy falling on 1 square centimetre of the plate is 5×10^{-6} ergs per sec. and the amount of energy per cubic centimetre of this radiation is 1.6×10^{-16} ergs.

According to Sir J.J. Thomson this value sets an upper limit to the amount of energy contained in one of the indivisible units mentioned above.

Reprinted from *Phys. Z.*, **18**, 121, translated from the German by D. ter Haar in *The Old Quantum Theory*, Pergamon, Oxford, 1967, pp. 167–173

On the Quantum Theory of Radiation

A. EINSTEIN

The formal similarity of the curve of the chromatic distribution of black-body radiation and the Maxwell velocity-distribution is too striking to be hidden for long. Indeed, already Wien in his important theoretical paper in which he derived his displacement law

$$\rho = \nu^3 f(\nu/T) \tag{1}$$

was led by this similarity to a further determination of the radiation formula. It is well known that he then found the formula

$$\rho = \alpha\nu^3 \; e^{-h\nu/kT}, \tag{2}$$

which is also nowadays accepted as being correct as a limiting law for large values of ν/T (Wien's radiation law). We know nowadays that no considerations based on classical mechanics and electrodynamics can give us a usable radiation formula, and that classical theory necessarily leads to the Rayleigh formula

$$\rho = \frac{k\alpha}{h}\, \nu^2 T\ . \tag{3}$$

As soon as Planck in his classical investigation based his radiation formula

$$\rho = \alpha\nu^3 \; \frac{1}{e^{h\nu/kT} - 1} \tag{4}$$

on the assumption of discrete elements of energy, from which very quickly quantum theory developed, it was natural that Wien's discussion which led to equation (2) became forgotten.

Recently[1†] I found a derivation of Planck's radiation formula which is based upon the basic assumption of quantum theory and which is related to Wien's

†The considerations given in that paper are repeated in the present one.

original considerations; in this derivation, the relationship between the Maxwell distribution and the chromatic black-body distribution plays a role. This derivation is of interest not only because it is simple, but especially because it seems to clarify somewhat the at present unexplained phenomena of emission and absorption of radiation by matter. I have shown, on the basis of a few assumptions about the emission and absorption of radiation by molecules, which are closely related to quantum theory, that molecules distributed in temperature equilibrium over states in a way which is compatible with quantum theory are in dynamic equilibrium with the Planck radiation. In this way, I deduced in a remarkably simple and general manner Planck's formula (4). It was a consequence of the condition that the distribution of the molecules over the states of their internal energy, which is required by quantum theory, must be established solely through the absorption and emission of radiation.

If the assumptions about the interaction between radiation and matter which we have introduced are essentially correct, they must, however, yield more than the correct statistical distribution of the internal energy of the molecules. In fact, in absorption and emission of radiation, momentum is transferred to the molecules; this entails that merely through the interaction of radiation and molecules the velocities of the molecules will acquire a certain distribution. This must clearly be the same velocity distribution as the one which the molecules attain through the action of their mutual collisions alone, that is, it must be the same as the Maxwell distribution. We must require that the average kinetic energy (per degree of freedom) which a molecule acquires in the Planck radiation field of temperature T is equal to $\frac{1}{2}kT$; this must be true independent of the nature of the molecules considered and independent of the frequencies of the light emitted or absorbed by them. In the present paper, we want to show that our simple hypotheses about the elementary processes of emission and absorption obtain another support.

In order to obtain the above-mentioned result we must, however, complete to some extent the hypotheses upon which our earlier work was based, as the earlier hypotheses were concerned only with the exchange of energy. The question arises: does the molecule receive an impulse when it absorbs or emits the energy ε? Let us, for instance, consider the emission from the point of view of classical electrodynamics. If a body emits the energy ε, it receives a recoil (momentum) ε/c if all of the radiation ε is emitted in the same direction. If, however, the emission takes place as an isotropic process, for instance, in the form of spherical waves, no recoil at all occurs. This alternative also plays a role in the quantum theory of radiation. When a molecule during a transition from one quantum-theoretically possible state to another absorbs or emits energy ε in the form of radiation, such an elementary process can be thought of either as being a partially or completely directed or as being a symmetrical (non-directional) process. *It now turns out that we arrive at a consistent theory only if we assume each elementary process to be completely directional.* This is the main result of the following considerations.

1. BASIC HYPOTHESIS OF QUANTUM THEORY CANONICAL DISTRIBUTION OVER STATES

According to quantum theory, a molecule of a given kind can take up - apart from its orientation and its translational motion - only a discrete set of states $Z_1, Z_2, \ldots, Z_n, \ldots$ with (internal) energies $\varepsilon_1, \varepsilon_2, \ldots, \varepsilon_n, \ldots$. If molecules of this kind form a gas of temperature T, the relative occurrence W_n of these states Z_n is given by the formula giving the canonical distribution of statistical mechanics:

$$W_n = p_n e^{-\varepsilon_n/kT} . \tag{5}$$

In this equation $k = R/N$ is the well-known Boltzmann constant, and p_n a number which is characteristic for the molecule and its *n*th quantum state and which is independent of *T*; it can be called the statistical "weight" of the state. One can derive equation (5) either from Boltzmann's principle or by purely thermodynamic means. Equation (5) expresses the greatest generalisation of Maxwell's velocity distribution law.

Recent important progress in quantum theory relates to the theoretical determination of quantum theoretically possible states Z_n and their weight p_n. For our considerations of the principles involved in radiation, we do not need a detailed determination of the quantum states.

2. HYPOTHESES ABOUT ENERGY EXCHANGE THROUGH RADIATION

Let Z_n and Z_m be two quantum-theoretically possible states of the gas molecule, and let their energies ε_n and ε_m satisfy the inequality $\varepsilon_m > \varepsilon_n$. Let the molecule be able to make a transition from the state Z_n to the state Z_m by absorbing radiative energy $\varepsilon_m - \varepsilon_n$; similarly let a transition from Z_m to Z_n be possible in which this radiative energy is emitted. Let the frequency of the radiation absorbed or emitted by the molecule in such transitions be ν; it is characteristic for the combination (m,n) of the indices.

We make a few hypotheses about the laws valid for this transition; these are obtained by using the relations known from classical theory for a Planck resonator, as the quantum-theoretical relations which are as yet unknown.

(a) *Spontaneous emission.*† It is well known that a vibrating Planck resonator emits according to Hertz energy independent of whether it is excited by an external field or not. Accordingly, let it be possible for a molecule to make without external stimulation a transition from the state Z_m to the state Z_n while emitting the radiation energy $\varepsilon_m - \varepsilon_n$ of frequency ν. Let the probability dW that this will in fact take place in the time interval dt be

$$dW = A_m^n \, dt , \tag{A}$$

where A_m^n denotes a constant which is characteristic for the combination of indices considered.

The statistical law assumed here corresponds to the law of a radioactive reaction, and the elementary process assumed here corresponds to a reaction in which only γ-rays are emitted. It is not necessary to assume that this process takes place instantaneously; it is only necessary that the time this process takes is negligible compared with the time during which the molecule is in the state Z_1, ...

†Einstein uses *Ausstrahlung* and *Einstrahlung* for spontaneous emission and induced radiation [D.t.H.].

(b) *Induced radiation processes.* If a Planck resonator is in a radiation field, the energy of the resonator can be changed by the transfer of energy from the electromagnetic field to the resonator; this energy can be positive or negative depending on the phases of the resonator and of the oscillating field. Accordingly we introduce the following quantum-theoretical hypothesis. Under the influence of a radiation density ρ of frequency ν a molecule can make a transition from the state Z_n to the state Z_m by absorbing the radiative energy $\varepsilon_m - \varepsilon_n$ and the probability law for this process is

$$dW = B_n^m \rho \, dt. \tag{B}$$

Similarly, a transition $Z_m \rightarrow Z_n$ may also be possible under the influence of the radiation; in this process the radiative energy $\varepsilon_m - \varepsilon_n$ will be freed according to the probability law

$$dW = B_m^n \rho \, dt. \tag{B'}$$

The B_n^m and B_m^n are constants. These two processes we shall call "changes in state, induced by radiation".

The question now arises: what is the momentum transferred to the molecule in these changes in state? Let us begin with the induced processes. If a radiation beam with a well-defined direction does work on a Planck resonator, the corresponding energy is taken from the beam. According to the law of conservation of momentum, this energy transfer corresponds also to a momentum transfer from the beam to the resonator. The resonator is thus subject to the action of a force in the direction of the beam. If the energy transferred is negative, the action of the force on the resonator is also in the opposite direction. This means clearly the following in the case of the quantum hypothesis. If through the irradiation by a beam of light a transition $Z_n \rightarrow Z_m$ is induced, the momentum $(\varepsilon_m - \varepsilon_n)/c$ is transferred to the molecule in the direction of propagation of the beam. In the induced transition $Z_m \rightarrow Z_n$ the transferred momentum has the same magnitude but is in the opposite direction. We assume that in the case where the molecule is simultaneously subjected to several radiation beams, the total energy $\varepsilon_m - \varepsilon_n$ of an elementary process is absorbed from or added to *one* of these beams, so that also in that case the momentum $(\varepsilon_m - \varepsilon_n)/c$ is transferred to the molecule.

In the case of a Planck resonator, when the energy is emitted through a spontaneous emission process, no momentum is transferred to the resonator, since according to classical theory the emission is in the form of a spherical wave. We have, however, already noted that we can only obtain a consistent quantum theory by assuming that the spontaneous emission process is also a directed one. In that case, in each spontaneous emission elementary process $(Z_m \rightarrow Z_n)$ momentum of magnitude $(\varepsilon_m - \varepsilon_n)/c$ is transferred to the molecule. If the molecule is isotropic, we must assume that all directions of emission are equally probable. If the molecule is not isotropic, we arrive at the same statement if the orientation changes in a random fashion in time. We must, of course, make a similar assumption for the statistical laws (B) and (B') for the induced processes, as otherwise the constants should depend on direction, but we can avoid this through the assumption of isotropy or pseudo-isotropy (through time-averaging) of the molecule.

3. DERIVATION OF THE PLANCK RADIATION LAW

We now ask for that radiation density ρ which must be present in order that the exchange of energy between radiation and molecules according to the statistical laws (A), (B), and (B') does not perturb the distribution (5) of the molecules. For this it is necessary and sufficient that on the average per unit time as many elementary processes of type (B) take place as of types (A) and (B') combined. This combination leads, because of (5), (A), (B), and (B'), to the following equation for the elementary processes corresponding to the index combination (m,n):

$$p_n e^{-\varepsilon_n/kT} B_n^m \rho = p_m e^{-\varepsilon_m/kT} (B_m^n \rho + A_m^n) .$$

If, furthermore, ρ will increase to infinity with T, as we shall assume, the following relation must exist between the constants B_n^m and B_m^n:

$$p_n\ B_n^m = p_m\ B_m^n \tag{6}$$

We then obtain from our equation the following condition for dynamic equilibrium:

$$\rho = \frac{A_m^n/B_m^n}{e^{(\varepsilon_m - \varepsilon_n)/kT} - 1} . \tag{7}$$

This is the temperature-dependence of the radiation density of the Planck law. From Wien's displacement law (1) it follows from this immediately that

$$\frac{A_m^n}{B_m^n} = \alpha\nu^3 , \tag{8}$$

and

$$\varepsilon_m - \varepsilon_n = h\nu , \tag{9}$$

where α and h are constants. To find the numerical value of the constant α, we should have an exact theory of the electrodynamic and mechanic processes; for the time being we must use the Rayleigh limit of high temperatures, for which the classical theory is valid as a limiting case.

Equation (9) is, of course, the second main hypothesis of Bohr's theory of spectra of which we can now state after Sommerfeld's and Epstein's extensions that it belongs to those parts of our science which are sure. It contains implicitly, as I have shown, also the photochemical equivalence rule.

4. METHOD OF CALCULATING THE MOTION OF MOLECULES IN THE RADIATION FIELD

We now turn to the investigation of the motions which our molecules execute under the influence of the radiation. To do this, we use a method which is well known from the theory of Brownian motion, and which I have used already many times for numerical calculations of motion in radiation. To simplify the calculations, we only perform them for the case where the motion takes place only in one direction, the X-direction of our system or coordinates. We shall moreover restrict ourselves to calculating the average value of the kinetic energy of the translational motion, and thus do not give the proof that these velocities v are distributed according to Maxwell's law. Let the mass M of the molecule be sufficiently large that we can neglect higher powers of v/c in comparison with lower ones; we can then apply ordinary mechanics to the molecule. Moreover, without any real loss of generality, we can perform our calculations as if the states with indices m and n were the only ones which the molecule can take on.

The momentum Mv of a molecule is changed in two ways in the short time τ. Although the radiation is the same in all directions, because of its motion the molecule will feel a force acting in the opposite direction of its motion which comes from the radiation. Let this force be Rv, where R is a constant to be evaluated later on. This force would bring the molecule to rest, if the irregularity of the action of the radiation did not have as a consequence that during the time τ a momentum Δ of varying sign and varying magnitude is transferred to the molecule; this unsystematic influence will against the earlier mentioned force maintain a certain motion of the molecule. At the end of the short time τ, which we are considering, the momentum of the molecule will have the value

$$Mv - Rv\tau + \Delta.$$

As the velocity distribution must remain the same in time, this quantity must have the same average absolute magnitude as Mv; therefore, the average squares of those two quantities, averaged over a long period or over a large number of molecules, must be equal to one another:

$$\overline{(Mv - Rv\tau + \Delta)^2} = \overline{(Mv)^2} .$$

As we have separately taken into account the systematic influence of v on the momentum of the molecule, we must neglect the average $\overline{\Delta v}$. Expanding the left-hand side of the equation, we get then

$$\overline{\Delta^2} = 2RM\overline{v^2}\tau . \tag{10}$$

The average $\overline{v^2}$ for our molecules, which is caused by radiation of temperature T through its interaction with the molecules must be equal to the average value $\overline{v^2}$, which according to the kinetic theory of gases a molecule in the gas would have according to the gas laws at the temperature T. Otherwise the presence of our molecules would disturb the thermodynamic equilibrium between black-body radiation and any gas of the same temperature. We must then have

$$\tfrac{1}{2}M\overline{v^2} = \tfrac{1}{2}kT . \tag{11}$$

Equation (10) thus becomes

$$\frac{\overline{\Delta^2}}{\tau} = 2RkT \; . \tag{12}$$

The investigation must now proceed as follows. For given radiation [$\rho(\nu)$] we can calculate Δ^2 and R with our hypotheses about the interaction between radiation and molecules. Inserting these results into (12), this equation must be satisfied identically if ρ as function of ν and T is expressed by the Planck equation (4).

5. CALCULATION OF R

Let a molecule of the kind considered move uniformly with velocity v along the X-axis of the system of coordinates K. We ask for the average momentumtum transferred by the radiation to the molecule per unit time. To be able to evaluate this, we must consider the radiation in a system of coordinates K' which is at rest relative to the molecule under consideration, because we have only formulated our hypotheses about emission and absorption for molecules at rest. The transformation to the system K' has often been given in the literature and especially accurately in Mosengeil's Berlin thesis.[2] For the sake of completeness, I shall, however, repeat the simple considerations.

In K the radiation is isotropic, that is, we have for the radiation per unit volume in a frequency range $d\nu$ and propagating in a direction within a given infinitesimal solid angle $d\kappa$:

$$\rho \, d\nu \, \frac{d\kappa}{4\pi} \, , \tag{13}$$

where ρ depends only on the frequency ν, but not on the direction. This particular radiation corresponds in the coordinate system K' to a particular radiation, which is also characterised by a frequency range $d\nu'$ and a certain solid angle $d\kappa'$. The volume density of this particular radiation is

$$\rho'(\nu',\phi')d\nu' \, \frac{d\kappa'}{4\pi} \, . \tag{13'}$$

This defines ρ'. It depends on the direction which is defined in the usual way by the angle ϕ' with the X'-axis and the angle ψ' between the projection in the Y'Z'-plane with the Y'-axis. These angles correspond to the angles ϕ and ψ which in a similar manner fix the direction of $d\kappa$ with respect to K.

First of all it is clear that the same transformation law must be valid bètween (13) and (13') as between the squares of the amplitude A^2 and A'^2 of a plane wave of the appropriate direction of propagation. Therefore in the approximation we want, we have

$$\frac{\rho'(\nu',\phi')d\nu' d\kappa'}{\rho(\nu)d\nu \, d\kappa} = 1 - 2\,\frac{v}{c}\cos\phi \, , \tag{14}$$

or

$$\rho'(\nu',\phi') = \rho(\nu)\,\frac{d\nu}{d\nu'}\,\frac{d\kappa}{d\kappa'}\left(1 - 2\,\frac{v}{c}\cos\phi\right). \tag{14'}$$

The theory of relativity further gives the following formulae, valid in the approximation needed here,

$$\nu' = \nu\left(1 - \frac{v}{c}\cos\phi\right), \tag{15}$$

$$\cos\phi' = \cos\phi - \frac{v}{c} + \frac{v}{c}\cos^2\phi\ , \tag{16}$$

$$\psi' = \psi\ . \tag{17}$$

From (15) in the same approximation it follows that

$$\nu = \nu'\left(1 + \frac{v}{c}\cos\phi'\right).$$

Therefore, also in the same approximation,

$$\rho(\nu) = \rho\left(\nu' + \frac{v}{c}\nu'\cos\phi'\right)\ ,$$

or

$$\rho(\nu) = \rho(\nu') + \frac{\partial\rho(\nu')}{\partial\nu}\frac{v}{c}\nu'\cos\phi'\ . \tag{18}$$

Furthermore from (15), (16), and (17) we have

$$\frac{d\nu}{d\nu'} = 1 + \frac{v}{c}\cos\phi'\ ,$$

$$\frac{d\kappa}{d\kappa'} = \frac{\sin\phi d\phi d\psi}{\sin\phi' d\phi' d\psi'} = \frac{d(\cos\phi)}{d(\cos\phi')} = 1 - 2\frac{v}{c}\cos\phi'\ .$$

Using these two relations and (18), we get from (14')

$$\rho(\nu',\phi') = \left[\rho(\nu) + \frac{v}{c}\nu'\cos\phi'\frac{\partial\rho(\nu)}{\partial\nu}\right]\left(1 - 3\frac{v}{c}\cos\phi'\right)\ . \tag{19}$$

From (19) and our hypothesis about the spontaneous emission and the induced processes of the molecule, we can easily calculate the average momentum transferred per unit time to the molecule. Before doing this we must, however, say something to justify the method used. One can object that the equations (14), (15), and (16) are based upon Maxwell's theory of the electromagnetic field which is incompatible with quantum theory. This objection, however, touches the form rather than the essence of the matter. Whatever the form of the theory of electromagnetic processes, surely in any case the Doppler principle and the aberration law will remain valid, and thus also equations (15) and (16). Furthermore, the validity of the energy relation (14) certainly extends beyond that of the wave theory; this transformation law is, for instance, also valid according to the theory of relativity for the energy density of a mass having an infinitesimal rest mass and moving with (quasi-) light-velocity. We can thus claim the validity of equation (19) for any theory of radiation.

The radiation corresponding to the spatial angle $d\kappa'$ will according to (B) lead per second to

$$B_n^m \rho'(\nu',\phi') \frac{d\kappa'}{4\pi}$$

induced elementary processes of the type $Z_n \to Z_m$, provided the molecule is brought back to the state Z_n immediately after each such elementary process. In reality, however, the time spent per second in the state Z_n is according to (5) equal to

$$\frac{1}{S} p_n e^{-\varepsilon_n/kT} ,$$

where we used the abbreviation

$$S = p_n e^{-\varepsilon_n/kT} + p_m e^{-\varepsilon_m/kT} . \tag{20}$$

In actual fact the number of these processes per second is thus

$$\frac{1}{S} p_n e^{-\varepsilon_n/kT} B_n^m \rho'(\nu',\phi') \frac{d\kappa'}{4\pi} .$$

In each process the momentum

$$\frac{\varepsilon_m - \varepsilon_n}{c} \cos\phi'$$

is transferred to the molecule in the direction of the positive X'-axis. Similarly, we find, using (B') that the corresponding number of induced elementary processes of the kind $Z_m \to Z_n$ per second is

$$\frac{1}{S} p_m e^{-\varepsilon_m/kT} B_m^n \rho'(\nu',\phi') \frac{d\kappa'}{4\pi} ,$$

and in each such elementary process the momentum

$$-\frac{\varepsilon_m - \varepsilon_n}{c} \cos\phi'$$

is transferred to the molecule. The total momentum transferred per unit time to the molecule through induced processes is thus, taking (6) and (9) into account,

$$\frac{h\nu'}{cS} p_n B_n^m (e^{-\varepsilon_n/kT} - e^{-\varepsilon_m/kT}) \int \rho'(\nu',\phi') \cos\phi' \frac{d\kappa'}{4\pi} ,$$

where the integration is over all elements of solid angle. Performing the integration we get, using (19), the value

$$-\frac{h\nu}{c^2 S} \left(\rho - \frac{1}{3}\nu \frac{\partial\rho}{\partial\nu} \right) p_n B_n^m (e^{-\varepsilon_n/kT} - e^{-\varepsilon_m/kT}) v .$$

Here we have denoted the frequency involved by ν (instead of ν').

This expression represents, however, the total average momentum transferred per unit time to a molecule moving with a velocity v; because it is clear that the spontaneous emission processes which take place without the action of radiation do not have a preferential direction, considered in the system K', so that they can on average not transfer any momentum to the molecule. We obtain thus as the final result of our considerations:

$$R = \frac{h\nu}{c^2 S}\left(\rho - \frac{1}{3}\nu\frac{\partial\rho}{\partial\nu}\right) p_n B_n^m e^{-\varepsilon_n/kT}\ (1 - e^{-h\nu/kT})\ . \tag{21}$$

6. CALCULATION OF $\overline{\Delta^2}$

It is much simpler to calculate the influence of the irregularity of the elementary processes on the mechanical behaviour of the molecule, as we can base this calculation on a molecule at rest in the approximation which we have used from the start.

Let some event lead to the transfer of a momentum λ in the X-direction to a molecule. Let this momentum have varying sign and varying magnitude in different cases, but let there be such a statistical law for λ that the average value of λ vanishes. Let now $\lambda_1, \lambda_2, \ldots$ be the values of the momentum in the X-direction transferred to the molecule through several, independently acting causes so that the resultant transfer of momentum Δ is given by

$$\Delta = \sum \lambda_\nu$$

As the average value $\overline{\lambda_\nu}$ vanishes for the separate λ_ν, we must have

$$\overline{\Delta^2} = \sum \overline{\lambda_\nu^2}\ . \tag{22}$$

If the averages λ_ν^2 of the separate momenta are equal to one another $(= \overline{\lambda^2})$, and if l is the total number of momentum transferring processes, we have the relation

$$\overline{\Delta^2} = l\overline{\lambda^2}\ . \tag{22a}$$

According to our hypothesis in each elementary process, induced or spontaneous, the momentum

$$\lambda = \frac{h\nu}{c}\cos\phi$$

is transferred to the molecule. Here ϕ is the angle between the X-axis and a direction chosen randomly. Therefore we have

$$\overline{\lambda^2} = \frac{1}{3}\left(\frac{h\nu}{c}\right)^2\ . \tag{23}$$

As we assume that we may take all elementary processes which take place to be independent of one another, we may apply (22a). In that case, l is the number of all elementary processes taking place during the time τ. This is twice the number of the number of induced processes $Z_n \rightarrow Z_m$ during the time τ. We have thus

$$l = \frac{2}{S} p_n B_n^m\ e^{-\varepsilon_n/kT}\ \rho\tau\ . \tag{24}$$

We get from (23), (24) and (22)

$$\frac{\overline{\Delta}^2}{\tau} = \frac{2}{3S}\left(\frac{h\nu}{c}\right)^2 p_n \, B_n^m \, e^{-\varepsilon_n/kT} \rho \tag{25}$$

7. RESULTS

To prove now that the momenta transferred by the radiation to the molecules in accordance with our hypotheses never disturb the thermodynamic equilibrium, we only need to substitute the values (25) and (21) for $\overline{\Delta^2}/\tau$ and R which we have calculated after we have used (4) to replace in (21) the quantity

$$\left(\rho - \frac{1}{3}\nu \, \frac{\partial\rho}{\partial\nu}\right)(1 - e^{-h\nu/kT})$$

by $\rho h\nu/3kT$. It then turns out immediately that our fundamental equation (12) is identically satisfied.

The considerations which are now finished give strong support for the hypotheses given in Section 2 about the interaction between matter and radiation through absorption and emission processes, that is, through spontaneous and induced radiation processes. I was led to these hypotheses by my attempt to postulate as simply as possible a quantum theoretical behaviour of the molecules which would be similar to the behaviour of a Planck resonator of the classical theory. I obtained then in a natural fashion from the general quantum assumption for matter the second Bohr rule (equation (9)) as well as the Planck radiation formula.

The most important result seems to me, however, to be the one about the momentum transferred to the molecule in spontaneous or induced radiation processes. If one of our assumptions about this momentum transfer is changed, this would lead to a violation of equation (12); it seems hardly possible to remain in agreement with this relation which is required by the theory of heat otherwise than on the basis of our assumptions. We can thus consider the following as rather certainly proved.

If a ray of light causes a molecule hit by it to absorb or emit through an elementary process an amount of energy $h\nu$ in the form of radiation (induced radiation process), the momentum $h\nu/c$ is always transferred to the molecule, and in such a way that the momentum is directed along the direction of propagation of the ray if the energy is absorbed, and directed in the opposite direction, if the energy is emitted. If the molecule is subjected to the action of several directed rays of light, always only one of them will partipate in an induced elementary process; this ray alone defines then the direction of the momentum transferred to the molecule.

If the molecule undergoes a loss of energy of magnitude $h\nu$ without external influence, by emitting this energy in the form of radiation (spontaneous emission), this process is also a directed one. There is no emission in spherical waves. The molecule suffers in the spontaneous elementary process a recoil of magnitude $h\nu/c$ in a direction which is in the present state of the theory determined only by "chance".

These properties of the elementary processes required by equation (12) make it seem practically unavoidable that one must construct an essentially quantum theoretical theory of radiation. The weakness of the theory lies, on the one hand, in the fact that it does not bring any nearer the connection with the wave theory and, on the other hand, in the fact that it leaves moment and direction of the elementary processes to "chance"; all the same, I have complete confidence in the reliability of the method used here.

Still one more general remark may be made here. Practically all theories of black-body radiation are based on a consideration of the interaction between radiation and molecules. However, in general one restricts oneself to considering energy-exchange, without taking momentum-exchange into account. One feels easily justified in this as the smallness of the momenta transferred by the radiation entails that these momenta are practically always in reality negligible compared to other processes causing a change in motion. However, for the theoretical discussion, these small actions must be considered to be completely as important as the obvious actions of the energy-exchange through radiation, as energy and momentum are closely connected; one can, therefore, consider a theory to be justified only when it is shown that according to it the momenta transferred by the radiation to the matter lead to such motion as is required by the theory of heat.

REFERENCES

1. Einstein A 1916 *Verh. Dtsch. Phys. Ges.* *18* 318

2. Von Mosengeil K 1907 *Ann. Physik* *22* 867

Reprinted from *Contemp. Phys.*, **7** (1965), 45–53

Take a Photon . . .

O. R. FRISCH
Cavendish Laboratory, Cambridge, UK

When light falls on an ideal half-silvered mirror half of it goes through, the other half is reflected. This is readily understood if we consider light as waves: we get a transmitted and a reflected wave, each having half the original intensity. But light is also a stream of photons, each carrying the energy amount $h\nu$. The frequency ν of the waves is not changed by the mirror, so the individual photons in both the transmitted and the reflected beam must have the original energy $h\nu$. One concludes that the photons are not split but go into one beam or the other, at random.

If we recombine the two beams in an interferometer (fig. 1) we usually get interference fringes; because of a slight misalignment (often intentional) the phase relation of the two waves as they become superimposed varies from place to place and we get alternate reinforcement and cancellation of the two waves.

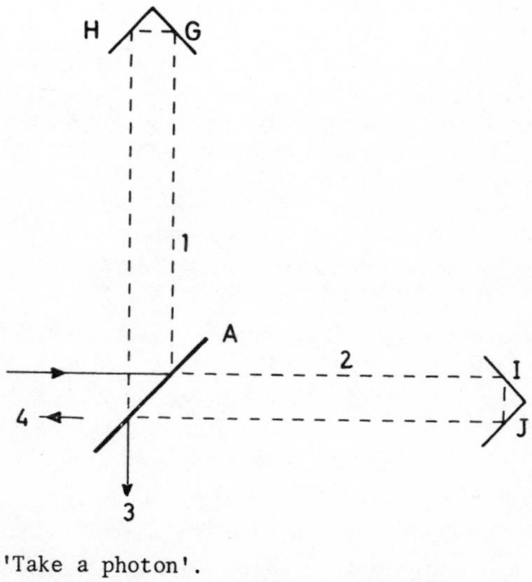

Fig.1 'Take a photon'.

But it is simpler to discuss an interferometer which is so adjusted that we get no fringes. For instance, if the two right-angle mirror pairs G-H and I-J are placed so that the intersection line of G and H is the exact mirror image (in the half-silvered mirror A) of the intersection line of I and J, then the light waves that pass through A on their return will cancel each other; all the light will go into beam 3 (the reflected beam), while beam 4 will have zero intensity.

But what happens to the *photons* in an interferometer? At first it was thought that interference occurred when two or more photons came together; but that was disproved when G.I. Taylor (1909) showed that interference fringes were formed just the same whether the light was strong or whether it was so weak that hardly ever two photons passed through the apparatus together. It follows that single photons can exhibit interference, that 'a photon can interfere with itself'. It would seem that something does travel along both paths in the interferometer even when only one photon is admitted; but what is it?

Such questions were discussed a good deal when photons were new, and similar questions arose out of the wave-particle duality of 'material' particles such as electrons. Some agreement has been reached on the way they should be answered, but the agreement is not unequivocal, and many of us are not sure what to tell our students. Indeed I am in two (or more) minds what to think, and for that reason I find it easiest to present the arguments in a dialogue between several characters. I have compelled them to be brief, but I'll try to elucidate what they say and sum up their findings at the end.

Jim: Take a photon...

Tom: How?

Jim: Well, take a weak light source and open a shutter long enough to let out one photon.

Tom: But you may get two, or none!

Bob: Is it single photons of visible light you want?

Jim: Yes.

Bob: Then I can help you; I have a generator for single photons of the sodium resonance line, $\lambda = 6 \cdot 10^{-5}$ cm. A beam of slow sodium atoms is crossed by a beam of yellow sodium light, which excites some of them. Those which emit a photon toward you are deflected on to a hot tungsten wire by their recoil, ionized and recorded by an electron multiplier. The output pulse tells us that a photon is on the way.

Tom: Won't it be gone before we observe the pulse?

Bob: There are lenses and mirrors to send the photon on a detour of 300 km, which gives you 1 millisecond notice; and there is a shutter which opens at the right time for an instant to let just that one photon pass through. The chance for another photon to arrive while the shutter is open is negligible.

Jim: Fine. So we take a photon...

Tom: Can you make sure you have one?

Bob: There is no need; my generator is quite reliable.

Tom: Still, one ought to be able to make sure there is a photon.

Jim: One could record it with a photomultiplier...

Tom: Not with any certainty!

Bob: True; photocathodes have at best about 30 percent efficiency. But with some semiconductors one can get close on 100 percent. There is some noise, but with deep cooling...

Tom: All right; let us say we have a perfect photon detector. So we can make a photon, know when it will come, and verify that it has come. But in verifying we kill it!

Jim: I know. Still, it seems I may at last say 'take a photon'. It behaves essentially like a particle: it starts from a point, it travels along a line, it...

Bill: Surely not! Light consists of waves; you can at best create a wave packet! And after travelling 300 km...

Bob: Let me give you the scale of my apparatus. My lens has a diameter a of one metre; over the distance $b = 300$ km the wave spreads by about $\lambda b/a$ which comes to 18 cm; the second lens is a little larger to allow for that, and it forms an image just as small as the original source, only a few wavelengths in size. The spread is...

Bill: Essentially nil, I agree. But where does the photon pass through those large lenses of yours?

Jim: I don't care; somewhere. Just let me take my photon from the focus of Bob's second lens. We can consider this as our photon source, and we know - we have 1 millisecond warning - when the photon is coming.

Tom: All right, we'll let you take a photon. What will you do with it?

Jim: I shall split it with a half-silvered mirror.

Bill: But that doesn't split the photon; it is either reflected or transmitted, the chances being half and half.

Jim: So the photon travels either in the direction 1 or 2?

Bill: Sure. If you were to place photon detectors in both beams, either one or the other would record the photon.

Jim: Good. Now please note that I have provided angle mirrors (fig.1) which cause both beams to return and to be recombined.

Bill: I see. You have built an interferometer similar to Michelson's. If your two distances are exactly alike then the reunited beam will go to the detector 3, not 4, if I've got the phase shifts right.

Jim: Correct. From that I conclude that the photon has been split, that it is present in both beams 1 and 2. There is no element of chance. We need both parts of the photon to obtain the interference that causes it to be recorded at 3 and not at 4.

Bill: But how do you account for the fact that of two photon detectors placed in 1 and 2, only one - at random - will record the photon?

Jim: It must be the detectors that introduce the randomness.

Bill: You mean that when one detector happens to report the photon, the other one is precluded from doing so?

Jim: Yes.

Bill: Even though the other half of the photon passes through it?

Jim: No. Surely the other half no longer exists, it must have been destroyed when the photon was spotted in the first beam. Only one photon was produced by Bob's source, and a photon can only be absorbed and detected once.

Tom: Isn't that the "reduction of the wave packet" that the theoreticians talk about?

Bob: I suppose so. But what does it mean? How can the observation of a photon in one place destroy the other half of the photon?

Bill: There is no split photon; there is only a wave, which indeed is split.

Bob: I must protest; my generator surely produces photons, one at a time. We agreed to that at the beginning.

Roy: Let me try to remember what I was taught. The wave associated with one photon is not real; it is merely a mathematical tool which allows us to compute the probability that a photon will be observed at a given place. The split wave is just a description of our knowledge that a single photon has entered our interferometer and has met a half-silvered mirror. Once we know that the photon has been observed in one of the beams the probability that it should be found in the other becomes nil.

Jim: Just as the chance for a horse to win a race becomes nil when another horse has won it?

Roy: A bit like that.

Tom: But if it is only a matter of probability that the photon is observed, might it not be missed by both detectors?

Roy: Let us assume the detector in beam 1 is nearer the splitting mirror than the one in beam 2. Then if it records the photon it modifies the wave - which only represents our knowledge - so that 2 has nothing to detect. But if detector 1 remains silent at the critical time, then the wave in beam 2 gets strengthened so that the photon is sure to be recorded there.

Tom: Which detector affects the wave if they are the same distance from the mirror, to within the length of the wavetrain? And anyhow, if the wave represents our knowledge it can only become modified by something that we come to know. What if we don't look at the first detector, but merely arrange for its signal to be recorded?

Roy: Then there will be an even chance - just as if the first detector wasn't there - that the second detector will report the photon.

Tom: Yes; but it will not report the photon in those cases where a later inspection of the first detector shows that it had, unknown to us at the time, recorded the photon. Does the present behaviour of the second detector then depend on the future state of our knowledge about the first one?

Roy: No. We must interpret knowledge in a wider way. When one of the detectors records the photon, then the way it went is 'known' though you and I may not know it.

Bob: This is getting ever more implausible: the knowledge stored in one box of electronics is said to affect a wave elsewhere - without signalling! - and so the behaviour of another box. Why not admit that the photon, on meeting the half-silvered mirror, takes a snap decision, at random, whether to go through or be deflected?

Jim: Because then you can't account for the interference. If you were sure that half the photons travel along each path you could block up one of the paths and merely halve the intensity recorded by detector 3. But you know that if you block one path you destroy the interference: you then observe as many photons in detector 4 as in 3, whereas with both paths open all the photons arrive at 3.

Bill: Wouldn't the wave account for the interference? There is both the wave and the photon. The wave gets split while the photon is either transmitted or reflected.

Jim: But if we block one path with a detector and find the photon has gone that way, then you still have a wave travelling along the other path; a futile little wave without a photon! Unless you 'reduce it', and then we are back to where we were.

Tom: Couldn't one spot the photon without absorbing it?

Bob: Certainly. For instance, a transparent block of mass M, thickness a and refractive index n will be displaced forward by the amount $s = a(n - 1)(h/Mc\lambda)$ when a photon of wavelength λ passes through it. That displacement is small, but...

Bill: I know. However, such a block causes a phase shift of $2\pi(n - 1)a/\lambda$ which will affect the interference and may destroy it.

Bob: Can't we choose our block so that the phase shift is $N.2\pi$ where N is an integer? Then it won't affect the interference.

Bill: Let me see. The displacement would come to $s = Nh/Mc$. If we measure to that accuracy we cannot know the momentum of our block to better than h/s (according to Heisenberg's uncertainty principle) or v to better than $h/sM = c/N$. If the block has the velocity v then the time the photon spends in the block is altered by the fraction $v/c = 1/N$, and so is the phase shift. So the phase is uncertain by 2π, and the interference is completely destroyed.

Roy: Look, all this has been threshed out in Copenhagen, in the 'thirties: if you spot the photon you ruin the phase.

Bob: I think I see a way around that. Let me suspend the half-silvered mirror so that we can measure its momentum perpendicular to its own plane...

Roy: That has all been disposed of. If you measure the momentum of the mirror to within h/λ, the momentum of a single photon, its position is uncertain by λ, and that ruins the interference.

Bob: No, wait, I have a new trick. I propose to suspend my mirror with a half-period equal to the time the photon takes to go out and back again along an arm of the interferometer. If it happened to deviate by +d from its equilibrium when the photon first arrived it will deviate by -d when the photon returns. Thus d has no effect on the phase, and the interference will not be ruined: it will happen just as if the mirror had been fixed in its equilibrium position?

Roy: Ingenious. How will you measure the momentum transfer?

Bob: Well, I just measure the momentum before the photon enters and again after it has been recorded at 3. The mirror will have been pushed one way if the photon was reflected at first and transmitted on the way back, and the other way if it followed the other path.

Roy: But in the latter case the push comes half a period later than in the first; and the final outcome is the same whether you push a pendulum in one direction or half a period later in the opposite. So your measurement does not tell you which path the photon has taken!

Bob: Ingenious. I fear you are right. So I must do one of my momentum measurements while the photon is in the interferometer, and that will spoil the interference.

Tom: Still, you have designed a means of observing where the photon is, without doing anything to it; does that not prove that the photon really is in one beam, and not split?

Jim: Einstein said something like that.

Roy: You have not really observed where the photon is, merely by suspending the mirror; you must measure its momentum.

Bob: But the momentum is in the mirror, and I can measure it or not, as I wish; surely that does nothing to the photon?

Roy: Yes, it does; it spoils the interference. All you have done is to share the clue about the position of the photon between the photon and the mirror so that it can be extracted from either of them. Your particular suspension ensures that this clue is automatically destroyed at the moment when the photon, by interference, gets directed to detector 3.

Tom: If the mirror had not been suspended in that way but just left floating, then the momentum transfer would have measured the photon's position?

Bob: Not necessarily. I think I could construct a mechanism by which you could make the mirror go to where the suspension would have taken it, if you so decide before the photon returns. All I need is...

Tom: We'll believe you. But once I have reflected a low-energy photon from the mirror so as to determine its velocity by Doppler shift, then the measurement is done?

Bob: Not necessarily. I might construct a system of mirrors to send back the low-energy photon and return its momentum to the mirror.

Tom: But when *is* the measurement done?

Bob: I think any reversible process can be reversed. But I would regard myself as beaten if you have let the system interact irreversibly with say, a semiconductor, a photographic grain, or a retina.

Tom: That makes sense. After all, to measure is to create information; and information is a state - in a machine or an organism - which extends from a certain time into the future. Irreversibility is the very essence of information.

Jim: But don't we sometimes obtain information without irreversible interaction? For instance, when the detector in beam 1 reports nothing we know that the photon is in beam 2.

Bob: Yes; but the detector has to be there, in beam 1; the possibility of an irreversible interaction is essential.

Jim: What if I place a piece of black paper into one beam? Then we have an irreversible interaction and we destroy the interference without getting any information in return.

Bob: Not necessarily; you could measure the temperature of the paper before and after!

Jim: But what if I don't?

Bob: Oh well, information can get lost if you are careless.

Jim: But, information aside, what does the photon do in my interferometer; does it get split, or doesn't it?

Roy: You musn't say 'information aside'. Quantum theory is about information. All it does is to tell you how to use available information to make the best possible predictions about future information.

Jim: You mean, about what is going to happen?

Bob: If you agree to use the word 'to happen' only for irreversible processes.

Tom: Surely something happens - in the everyday sense - to the photon inside the interferometer; so quantum theory must be incomplete.

Roy: I don't feel that. Quantum theory is logically consistent, and it allows you to make all the predictions that you can test. Photons and waves are models that allow you to use your imagination instead of using the full theory, but they cannot completely replace it.

Bill: Couldn't one have a model that covers both photons and waves? Something more complex, perhaps multidimensional, of which our present concepts are merely flat projections?

Roy: Plato's cave. Well, produce such a model, and we'll discuss it next time.

Jim: But there are some worse difficulties which today we haven't even touched on. Take two photons...

Here we must break off the discussion and see what has been achieved. First we must admit that many of Bob's ambitious gadgets will never be built. That need not worry us; the use of thought experiments ('Gedanken-Experimente') in physical arguments has a long history and is generally accepted. Of course a thought experiment may be faulty if it contains an essential snag that has been overlooked by its designer. (A famous example was a thought experiment designed by Einstein and refuted by Bohr in 1930). I have allowed Bob to explain his photon generator in some detail because photons are emitted at random, and some people have suggested that any thought experiment must be faulty if it assumes that we 'take a photon'. I think Bob has dispelled that idea. He cannot produce a photon at a specified time; but if Jim is willing to wait he will get one photon at a time and will know (a millisecond in advance) when it will arrive.

After that, the discussion turns to the question, what happens to the photon on striking the half-silvered mirror? Jim believes that it is split; Tom doubts it; Bill suggests that merely the wave is split. None of the attempted ways of visualizing what happens inside an interferometer appears satisfactory. The most common way is the one put forward by Roy, that the wave - which is split - determines the probability of the photon being intercepted in either beam 1 or 2. But then one has to assume that if the photon is found in, say, beam 1 the wave packet which represented the probability of finding it in beam 2 is thereby reduced to zero. This "reduction of the wave packet" is not a physical process in the ordinary sense; it happens instantly, however far the two wave packets may be apart at the time when the photon is found in one of them.

Roy suggests that we should take the wave merely as a representation of what we know about the photon. Then the reduction process seems easier to understand; the wave packet becomes comparable to a list of betting odds which drop to zero as soon as a different contestant is known to have won the race. But it is not just a matter of our knowledge (say, that one detector has reported the photon) affecting our belief (concerning the chance of the other detector reporting it). We can place ideal photon detectors into both beams and record their signals; afterwards we can verify that every photon admitted was recorded by either one or the other of the two detectors. If we use the idea of a wave that is split we must assume collusion between the detectors (which is what the "reduction of the wave packet" amounts to).

Yet we cannot simply assume - as Tom suggests - that each photon is either transmitted or reflected; if that were so we could not account for the interference, that is, for the fact that all the photons emerge in beam 3 and not 4, provided both paths are kept open. If we want to visualize what causes the interference, we must think of a wave that is split by the half-silvered mirror and explores both paths.

Those two descriptions - a wave that is split and capable of interference, and a photon that is at random either transmitted or reflected - are called *complementary* aspects of the same thing, according to Niels Bohr. According to that view it is up to the physicist to choose that description which is appropriate to a given experimental situation. For the interferometer the wave picture is appropriate; if detectors are used to monitor the two beams (or even one of them) then the photon picture is appropriate. But there are intermediate situations (not discussed here) when neither picture is adequate, and then the mathematical theory must be wheeled into position.

The discussion then turned to the question whether the photon might not be spotted in one of the two beams without being absorbed. That is indeed possible (at least in the realm of thought experiments). For instance, a photon entering a block of glass changes its momentum from h/λ to $h/n\lambda$; the difference in momentum $(n-1)h/\lambda n$ is given to the glass block and causes it to move forward with the speed $(n-1)h/\lambda Mn$. After the time an/c the block stops again as the photon emerges from it; it has thus travelled the distance $(n-1)ah/cM\lambda$ (as Bob says).

But even though we have not absorbed the photon we have lost the chance of observing interference. Heisenberg's uncertainty principle tell us that the speed of the glass block will be uncertain by $\Delta v = \Delta p/M = h/Ms$ if we know its position to the accuracy s. That speed affects the time the photon spends in the block and hence the phase shift by just enough to make it uncertain whether the photon will go into beam 3 or 4. So once again, if we locate the photon we no longer observe the phenomenon - interference - which caused Jim to believe that the photon gets split.

At that point the inventive Bob comes up with a variant of the movable-mirror idea. His special suspension fails to allow him - as he first thought - to spot the photon and still to retain the interference. But it gives him liberty to choose: he can measure the momentum of the mirror (say by reflecting a low-energy - i.e. long-wave - photon from it and measuring the change, due to Doppler effect, of its wavelength) before the photon has returned from one of the two angle mirrors, and that will tell him which way it went; or, if he does nothing, the mirror will swing in such a way that the interference is not disturbed. This is really a model of the EPR paradox (Einstein, Podolski and Rosen 1935), namely the fact (demonstrated mathematically in that paper) that one can create situations in which one has the choice of measuring either the momentum or the position of a given particle "without physically affecting that particle". But that final clause (in inverted commas) doesn't really create a paradox once we have accepted that "the same thing" - a photon that has met a half-silvered mirror - requires two totally different descriptions, depending on whether we have arranged to locate the photon or to observe interference (Bohr 1935). Again, there is nothing new in Bob's suspended mirror, merely a concrete example of something that is hard to think about. More severe forms of the EPR paradox exist (Furry, Dicke and Wittke, Frisch) and Jim wanted to talk about them, but I had to cut him off.

Perhaps the most important conclusion arises out of the insistence of Bob that "any reversible process can be reversed", given enough ingenuity. The conclusion is that a measurement is not done until some irreversible process has taken place, such as an interaction with a grain in a photographic emulsion. Let me repeat what Tom says: "To measure is to create information, which is a state - in a machine or an organism - which extends from a certain time into the future". The intimate connection between information and irreversibility has indeed been stressed by L. Brillouin.

The main thing, as Roy points out, is that we must not ask what a quantum system does between observations. Quantum theory tells us how to use available information to make predictions about future information, and there are reasons to think (J. von Neumann) that they are the best possible predictions. I still feel a bit uneasy because I see no clear way of drawing the line between irreversible interactions which create (at least potentially) information and are to that extent 'real', and those interactions - like that between a photon and a half-silvered mirror - which don't create information and are 'unreal', demanding different descriptions in different circumstances.

As to Tom's last suggestion, of a possible model of which our present concepts - such as waves and photons - are merely shadows like those in Plato's cave, I have some sympathy with it; but no such model appears to be in sight.

* * * * * *

The dialogue that forms the centre-piece was written for the 70th birthday (on 15th September 1964) of Professor Oskar Klein, Stockholm, one of the pioneers of quantum theory, whose name is familiar from the Klein Paradox, the Klein-Gordon Wave equation and the Klein-Nishina formula.

REFERENCES

Bohr N 1935 *Phys. Rev.* *48* 696

Bohr N 1951 in A. Schilpp, *Einstein: Philosopher-Scientist.*

Brillouin L 1956, *Science and Information Theory* (Academic Press).

Dicke R H and Wittke J P 1960, *Introduction to Quantum Mechanics* (Addison-Wesley), see p.120.

Einstein A, Podolsky B and Rosen N 1935 *Phys. Rev.* *47* 777.

Frisch O R 1964, in M Bunge, *The Critical Approach to Science and Philosophy* (Macmillan) (a celebration volume for Karl Popper's 60th birthday).

Furry W H 1936 *Phys. Rev.* *49* 393 476.

Taylor G I 1909 *Proc. Camb. Phil. Soc.* *15* 114.

Reprinted from *Phys. Rev.*, **72** (1947), 241–243

Fine Structure of the Hydrogen Atom by a Microwave Method*,**

W. E. LAMB, Jr. and R. C. RETHERFORD
Columbia Radiation Laboratory, Department of Physics, Columbia University, New York, NY, USA

The spectrum of the simplest atom, hydrogen, has a fine structure[1] which according to the Dirac wave equation for an electron moving in a Coulomb field is due to the combined effects of relativistic variation of mass with velocity and spin-orbit coupling. It has been considered one of the great triumphs of Dirac's theory that it gave the "right" fine structure of the energy levels. However, the experimental attempts to obtain a really detailed confirmation through a study of the Balmer lines have been frustrated by the large Doppler effect of the lines in comparison to the small splitting of the lower or n = 2 states. The various spectroscopic workers have alternated between finding confirmation[2] of the theory and discrepancies[3] of as much as eight percent. More accurate information would clearly provide a delicate test of the form of the correct relativistic wave equation, as well as information on the possibility of line shifts due to coupling of the atom with the radiation field and clues to the nature of any non-Coulombic interaction between the elementary particles: electron and proton.

The calculated separation between the levels $2^2P_{\frac{1}{2}}$ and $2^2P_{3/2}$ is 0.365 cm^{-1} and corresponds to a wavelength of 2.74 cm. The great wartime advances in microwave techniques in the vicinity of three centimeters wavelength make possible the use of new physical tools for a study of the n = 2 fine structure states of the hydrogen atom. A little consideration shows that it would be exceedingly difficult to detect the direct absorption of radiofrequency radiation by excited H atoms in a gas discharge because of their small population and the high background absorption due to electrons. Instead, we have found a method depending on a novel property of the $2^2S_{\frac{1}{2}}$ level. According to the Dirac theory, this state exactly coincides in energy with the $2^2P_{\frac{1}{2}}$ state which is the lower of the two P states. The S state in the absence of external electric fields is metastable. The radiative transition to the ground state $1^2S_{\frac{1}{2}}$ is forbidden by the selection rule $\Delta L = \pm 1$. Calculations of Breit and Teller[4] have shown that the most probable decay mechanism is double quantum emission with a lifetime of 1/7 second. This is to be contrasted with a

*Publication assisted by the Ernest Kempton Adams Fund for Physical Research of Columbia University, New York, U.S.A.

**Work supported by the Signal Corps under contract number W 36-039 sc-32003.

lifetime of only 1.6 x 10^{-9} second for the non-metastable 2^2P states. The metastability is very much reduced in the presence of external electric fields[5] owing to Stark effect mixing of the S and P levels with resultant rapid decay of the combined state. If for any reason, the $2^2S_{\frac{1}{2}}$ level does not exactly coincide with the $2^2P_{\frac{1}{2}}$ level, the vulnerability of the state to external fields will be reduced. Such a removal of the accidental degeneracy may arise from any defect in the theory or may be brought about by the Zeeman splitting of the levels in an external magnetic field.

In brief, the experimental arrangement used is the following: Molecular hydrogen is thermally dissociated in a tungsten oven, and a jet of atoms emerges from a slit to be cross-bombarded by an electron stream. About one part in a hundred million of the atoms is thereby excited to the metastable $2^2S_{\frac{1}{2}}$ state. The metastable atoms (with a small recoil deflection) move on out of the bombardment region and are detected by the process of electron ejection from a metal target. The electron current is measured with an FP-54 electrometer tube and a sensitive galvanometer.

If the beam of metastable atoms is subjected to any perturbing fields which cause a transition to any of the 2^2P states, the atoms will decay while moving through a very small distance. As a result, the beam current will decrease, since the detector does not respond to atoms in the ground state. Such a transition may be induced by the application to the beam of a static electric field somewhere between source and detector. Transitions may also be induced by radiofrequency radiation for which $h\nu$ corresponds to the energy difference between one of the Zeeman components of $2^2S_{\frac{1}{2}}$ and any component of either $2^2P_{\frac{1}{2}}$ or $2^2P_{3/2}$. Such measurements provide a precise method for the location of the $2^2S_{\frac{1}{2}}$ state relative to the P states, as well as the distance between the latter states.

We have observed an electrometer current of the order of 10^{-14} ampere which must be ascribed to metastable hydrogen atoms. The strong quenching effect of static electric fields has been observed, and the voltage gradient necessary for this has a reasonable dependence on magnetic field strength.

We have also observed the decrease in the beam of metastable atoms caused by microwaves in the wavelength range 2.4 to 18.5 cm in various magnetic fields. In the measurements, the frequency of the r-f is fixed, and the change in the galvanometer current due to interruption of the r-f is determined as a function of magnetic field strength. A typical curve of quenching *versus* magnetic field is shown in Fig. 1. We have plotted in Fig. 2 the resonance magnetic fields for various frequencies in the vicinity of 10,000 Mc/sec. The theoretically calculated curves for the Zeeman effect are drawn as solid curves, while for comparison with the observed points, the calculated curves have been shifted downward by 1000 Mc/sec (broken curves). The results indicate clearly that, contrary to theory but in essential agreement with Pasternack's hypothesis[3], the $2^2S_{\frac{1}{2}}$ state is higher than the $2^2P_{\frac{1}{2}}$ by about 1000 Mc/sec.($0.033\ cm^{-1}$) or about 9 percent of the spin relativity doublet separation. The lower frequency transitions $^2S_{\frac{1}{2}}(m = \frac{1}{2}) \rightarrow {}^2P_{\frac{1}{2}}(m = \pm \frac{1}{2})$ have also been observed and agree well with such a shift of the $^2S_{\frac{1}{2}}$ level. With the present precision, we have not yet detected any discrepancy between the Dirac theory and the doublet separation of the P levels. (According to most of the imaginable theoretical explanations of the shift, the doublet separation would not be affected as much as the relative location of the S and P states.) With proposed refinements in sensitivity, magnetic field homogeneity, and calibration, it is hoped to locate the S level with respect to each P level to an accuracy of at least ten Mc/sec. By addition of these frequencies and assumption of the theoretical formula $\Delta\nu = 1/16\ \alpha^2 R$ for the doublet separation, it should

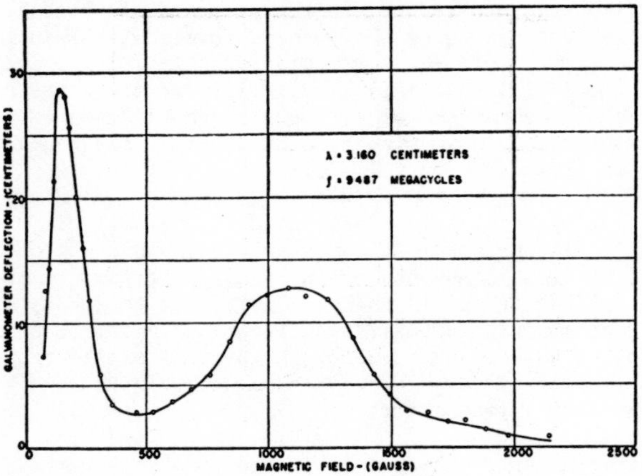

Fig. 1. A typical plot of galvanometer deflection due to interruption of the microwave radiation as a function of magnetic field. The magnetic field was calibrated with a flip coil and may be subject to some error which can be largely eliminated in a more refined apparatus. The width of the curves is probably due to the following causes: (1) the radiative line width of about 100 Mc/sec. of the ^2P states, (2) hyperfine splitting of the ^2S state which amounts to about 88 Mc/sec., (3) the use of an excessive intensity of radiation which gives increased absorption in the wings of the lines, and (4) inhomogeneity of the magnetic field. No transitions from the state $2^2S_{\frac{1}{2}}(m = -\frac{1}{2})$ have been observed, but atoms in this state may be quenched by stray electric fields because of the more nearly exact degeneracy with the Zeeman pattern of the ^2P states.

be possible to measure the square of the fine structure constant times the Rydberg frequency to an accuracy of 0.1 percent.

By a slight extension of the method, it is hoped to determine the hyperfine structure of the $2^2S_{\frac{1}{2}}$ state. All of these measurements will be repeated for deuterium and other hydrogen-like atoms.

A paper giving a fuller account of the experimental and theoretical details of the method is being prepared, and this will contain later and more accurate data.

The experiments described here were discussed at the Conference on the Foundations of Quantum Mechanics held at Shelter Island on June 1-3, 1947 which was sponsored by the National Academy of Sciences.

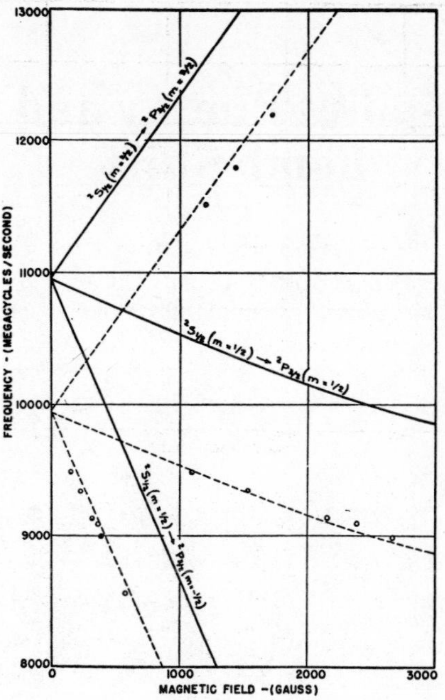

Fig. 2. Experimental values for resonance magnetic fields for various frequencies are shown by circles. The solid curves show three of the theoretically expected variations, and the broken curves are obtained by shifting these down by 1000 Mc/sec. This is done merely for the sake of comparison, and it is not implied that this would represent a "best fit". The plot covers only a small range of the frequency and magnetic field scale covered by our data, but a complete plot would not show up clearly on a small scale, and the shift indicated by the remainder of the data is quite compatible with a shift of 1000 Mc.

REFERENCES

1. For a convenient account, see White H E *Introduction to Atomic Spectra* (McGraw-Hill, New York, 1934), Chapter 8.

2. Drinkwater J W, Richardson O and Williams W E 1940 *Proc. Roy. Soc. 174* 164

3. Houston W V 1937 *Phys. Rev. 51* 446; Williams R C 1938 *Phys. Rev. 54* 558; Pasternack S 1938 *Phys. Rev. 54* 1113, has analyzed these results in terms of an upward shift of the S level by about 0.03 cm^{-1}.

4. Bethe H A in *Handbuch der Physik, Vol. 24/1,* §43

5. Breit G and Teller E 1940 *Astrophys. J. 91* 215

Reprinted from *Am. J. Phys.*, **34** (1966), 516–518

Zero-point Energy and the Lamb Shift

E. A. POWER
Mathematics Department, University College, London, UK and National Bureau of Standards, Boulder, Colorado, USA

A suggestion by Feynman that the Lamb-shift energy can be obtained from the changes in zero-point energy due to the presence of the atoms involved is worked through in detail for the Bethe nonrelativistic contribution to the energy shift. The self energy of a free electron is obtained in the same way and is equivalent to the state independent shift due to the $e^2A^2/2mc^2$ term in the normal perturbation methods.

INTRODUCTION

It has been demonstrated by Welton[1] how the zero-point fluctuations of the electromagnetic field give rise to an energy shift for electrons moving in the Coulomb field of a nucleus and in the fluctuating field. He assumed that the high frequency, high relative to the natural frequencies in the bound states involved, components of the field are followed by the electron according to Newton's law and the Lorentz force. It follows that there is a mean-square deviation from the orbits following the nuclear Coulomb field proportioned to k^{-3} (k being the magnitude of the wave-vector of the fluctuating field) for the mode k, from which there follows a change in energy proportional to $\int(dk/k)$ in Bethe's[2] Lamb shift calculation with cutoffs suitably chosen. At low frequencies, the cutoff is at the natural frequencies in the orbit while at high frequency the choice is dictated by the matching to the relativistic part where the electron can be considered free. The exact value does not matter but the order of magnitude which lets the nonrelativistic contribution to be the bulk of the total shift is that corresponding to the electron rest mass $k_{max} \sim mc/h$.

It has been suggested by Feynman[3] that the Lamb shift can be obtained from the changes in the zero-point energy due to the frequency changes implied by a weak perturbing background of atoms acting as a refracting medium. It is the purpose of this note to work this idea through and demonstrate the equivalence to Welton's and Bethe's results.

CALCULATION

Consider a box containing N hydrogen atoms in a state l and let the normal modes for standing waves have wavelength λ (<< linear size of the box). Then, if $\nu(l)$ is the frequency of such waves,

$$\nu(l)\lambda = c' , \tag{1}$$

where c' is the speed of light in the medium. So

$$c' = c/n(l) , \tag{2}$$

where $n(l)$ is the refractive index.

If N is small the refractive index is given in terms of the scattering amplitude β by the classical relation[4]

$$\frac{2\pi}{\lambda}(n - 1) = \beta\lambda N . \tag{3}$$

The zero-point energy is then

$$\begin{aligned} E_l &= \tfrac{1}{2}\sum h\nu(l) \\ &= \tfrac{1}{2}hc\sum_{\lambda}\frac{1}{\lambda}\frac{1}{n(l)} \end{aligned} \tag{4}$$

The differences between the energies for say the 2p and 2s levels of hydrogen is thus

$$\begin{aligned} E_{2s} - E_{2p} = \tfrac{1}{2}hc\sum_{\lambda}\frac{1}{\lambda}\Bigg[1\Big/\left(1 + \frac{1}{2\pi}\beta(2s)\lambda^2N\right) \\ - 1\Big/\left(1 + \frac{1}{2\pi}\beta(2p)\lambda^2N\right)\Bigg] \end{aligned} \tag{5}$$

and the difference per atom is

$$\frac{E_{2s} - E_{2p}}{N} \approx -\tfrac{1}{2}hc\sum_{\lambda}\frac{\lambda}{2\pi}\{\beta(2s) - \beta(2p)\} \tag{6}$$

for small N. From the Kramers-Heisenberg dispersion formula[5]

$$\beta(l) = k^2\sum_m \frac{|\langle l|\mu|m\rangle|^2 2(E_m - E_l)}{(\hbar ck)^2 - (E_m - E_l)^2} \tag{7}$$

where μ is the dipole moment of the system and only dipole transitions are considered. The states $|m\rangle$ are those accessible from $|l\rangle$ by such transitions.

If one follows Welton in considering the bulk of the energy shift as arising from those fluctuations with k larger than the natural frequencies within the atom then

$$\frac{E_{2s} - E_{2p}}{N}$$

$$\simeq + \pi \sum_k \frac{1}{\hbar ck} \Bigg\{ \sum_m |\langle 2s|\mu|m\rangle|^2 2(E_m - E_{2s})$$

$$- \sum_{m'} |\langle 2p|\mu|m'\rangle|^2 2(E_{m'} - E_{2p})$$

$$+ \frac{1}{(\hbar ck)^2} \sum_m |\langle 2s|\mu|m\rangle|^2 2(E_m - E_{2s})^3$$

$$- \frac{1}{(\hbar ck)^2} \sum_{m'} |\langle 2p|\mu|m'\rangle|^2 2(E_{m'} - E_{2p})^3 \Bigg\} . \tag{8}$$

The atomic sums over m and m' can all be evaluated by the closure rules

$$\sum_m |\langle l|\mu|m\rangle|^2 (E_m - E_l) = \frac{e^2\hbar^2}{2m}$$

and

$$\sum_m |\langle l|\mu|m\rangle|^2 (E_m - E_l)^3 = \frac{e^2\hbar^4}{6m^2} \langle l|\nabla^2 V|l\rangle . \tag{9}$$

The leading terms in eq.(8) cancel as they are independent of l, and so finally

$$W_{2s,2p} = - \frac{\pi}{(\hbar c)^3} \sum_k \frac{1}{k^3} \frac{e^2 h^4}{3m^2} (- 4\pi e^2) |\psi_{2s}(0)|^2$$

$$= \frac{4}{3}\left(\frac{e^2}{\hbar c}\right)^2 \left(\frac{\hbar}{mc}\right)^2 \hbar c |\psi_{2s}(0)|^2 \int \frac{dk}{k} \tag{10}$$

and is the Welton result with the cutoffs mentioned in the introduction,

$$= \frac{4}{3} \left(\frac{e^2}{\hbar c}\right)^2 \left(\frac{\hbar}{mc}\right)^2 \hbar c |\psi_{2s}(0)|^2 \log \left(\frac{mc}{\hbar k_0}\right) . \tag{11}$$

The expansion involved in eq.(8) is not essential to obtain the theoretical Lamb shift energy. If the sum over k value is made exactly from zero to Λ the result is

$$W_{2s,2p} \approx - \frac{\pi}{\hbar c} \int_0^\Lambda k^3 \Bigg\{ \sum_m \frac{|\langle 2s|\mu|m\rangle|^2 2(E_m - E_{2s})}{k^2 - [(E_m - E_{2s})/\hbar c]^2}$$

$$- \sum_{m'} \frac{|\langle 2p|\mu|m'\rangle|^2 2(E_{m'} - E_{2p})}{k^2 - [(E_{m'} - E_{2p})/\hbar c]^2} \Bigg\} \frac{dk 8\pi}{(2\pi)^3} \tag{12}$$

The integrals are principal values at any pole; this is always the prescription for an energy-shift calculation. The leading terms cancel because of the first sum rule of eq.(9) and

$$W_{2s,2p} \approx - \frac{1}{\pi(\hbar c)^3}\left\{\sum_m |\langle 2s|\mu|m\rangle|^2 (E_m - E_{2s})^3 \times \tfrac{1}{2}\log\left|\frac{\Lambda^2 - [(E_m - E_{2s})/\hbar c]^2}{[(E_m - E_{2s})/\hbar c]^2}\right| - \sum_m |\langle 2p|\mu|m\rangle|^2 (E_{m'} - E_{2p})^3 \times \tfrac{1}{2}\log\left|\frac{\Lambda^2 - [(E_{m'} - E_{2p})/\hbar c]^2}{[(E_{m'} - E_{2p})/\hbar c]^2}\right|\right\} . \tag{13}$$

The logarithm has a weak dependence on E_m and, in exact analogy to Bethe's calculation, it is convenient to define an average energy $\bar{E}$ through the equation

$$\sum_m |\langle 2s|\mu|m\rangle|^2 (E_m - E_{2s})^3 \log\frac{\Lambda\hbar c}{\bar{E}} = \sum_m |\langle 2s|\mu|m\rangle|^2 (E_m - E_{2s})^3 \times \tfrac{1}{2}\log\left|\frac{\Lambda^2 - [(E_m - E_{2s})/\hbar c]^2}{[(E_m - E_{2s})/\hbar c]^2}\right| , \tag{14}$$

so

$$W_{2s,2p} \approx - \frac{1}{\pi hc}\sum_m |\langle 2s|\mu|m\rangle|^2 \times (E_m - E_{2s})^3 \log\frac{\Lambda\hbar c}{\bar{E}} \tag{15}$$

and the second sum rule of eqn.(9) gives

$$W_{2s,2p} \approx \frac{4}{3}\left(\frac{e^2}{\hbar c}\right)^2\left(\frac{\hbar}{mc}\right)^2 \hbar c|\psi_{2s}(0)|^2 \log\frac{\Lambda\hbar c}{\bar{E}} , \tag{16}$$

which is exactly the usual nonrelativistic contribution with however $\bar{E}$ defined by eqn.(14). With $\Lambda \gg \bar{E}/\hbar c$ this is equivalent to the normal perturbation theory result.

FREE ELECTRON

The self energy of free electrons can also be computed by this method. The difference in zero-point energy with N electrons in the box and when empty is

$$\Delta E = -\tfrac{1}{2}hc\sum_{\lambda}\frac{1}{\lambda}\left(\frac{n-1}{n}\right), \tag{17}$$

where n is given by the Lorentz-Lorentz law

$$n - 1 = 2\pi\,\frac{Ne^2}{m(\omega_o^2 - \omega^2)} \tag{18}$$

which, for free electrons, is

$$n - 1 = -\,(2\pi Ne^2/m\omega^2)\ . \tag{19}$$

This corresponds to $\beta = r_o = e^2/mc^2$ in eq.(3). Hence the energy change is, for N small,

$$\Delta E = \tfrac{1}{2}\hbar c\sum_{\lambda}\frac{1}{\lambda}\,\frac{e^2\lambda^2}{mc^2} = \frac{\hbar e^2}{\pi mc}\int kdk \tag{20}$$

and is quadratically divergent for large k. This is precisely the self-energy computed by perturbation theory on the $(e^2/2mc^2)A^2$ term in the conventional nonrelativistic Hamiltonian of quantum electrodynamics[6]; it is independent of the state of the electron and thus changes the datum point for measuring energies and is usually ignored.[7]

1. Welton T A 1948 *Phys. Rev.* *74* 1157
2. Bethe H A 1947 *Phys. Rev.* *72* 339
3. Feynman R P *Solvay Institute Proceedings* (Interscience Publishers, Inc., New York, 1961) p.76
4. For example, Ditchburn R W *Light* (Interscience Publishers, Inc., New York, 1953) p.475
5. Kramers H A and Heisenberg W 1925 *Z. Physik 31* 681
6. For example, Power E A *Introductory Quantum Electrodynamics* (Longmans Green and Co. Ltd., London, 1964) p.119
7. For example, Schweber S S *Introduction to Quantum Field Theory* (Row, Peterson and Co., New York, 1961) p.525 and Kroll N *Quantum Optics and Electronics* (Gordon and Breach Science Publishers, Inc., New York, 1964) p.44

Reprinted from *Phys. Lett.*, **24a** (1967), 766–767

Interference Effects at the Single Photon Level*

R. L. PFLEEGOR and L. MANDEL
Department of Physics and Astronomy,
University of Rochester, Rochester, NY 14627, USA

Interference effects have been demonstrated in the superposition of two light beams from two independent lasers, under conditions where the intensity was so low that one photon was absorbed, with high probability, before the next one was emitted by either one of the sources.

Since the experimental demonstration of interference effects produced by the superposition of two independent light beams [1], the question has been debated whether the effect is to be regarded as evidence for the interference of photons from one beam with photons of the other beam [2-4]. On the face of it, the observations appeared to contradict a well-known remark of Dirac [5] that "...each photon interferes only with itself. Interference between different photons never occurs".

It is the purpose of this communication to report the results of further experiments which show that the effect cannot reasonably be described as an interference between photons of one beam and photons of the other. On the contrary, it appears that Dirac's statement, in a sense, is just as applicable to the foregoing experimental situation as to conventional interferometry.

We recall that, if $\Delta\nu$ is the overall frequency spread of the two light beams, assumed to be spatially coherent over the surface S on which the interference pattern is to be received, then any interference fringes will be expected to remain stationary only over a time interval $T < 1/\Delta\nu$ [6]. There is therefore an upper limit to the 'exposure time' in which the fringes may be examined. Now let the light intensities of the two beams be reduced until the mean rates r_1 and r_2 at which photons are striking the surface S satisfy $r_1 \ll 1/\tau$ and $r_2 \ll 1/\tau$, where τ is the transit time through the interferometer. Under these conditions it can be said that one photon is absorbed at S before the next one is emitted by one or the other source, with high probability. If interference fringes are formed under these conditions they cannot easily be described as an interference between two independent photons, but must be associated with the detection of *each photon*.

*This work was supported by the Air Force Office of Scientific Research, Office of Aerospace Research.

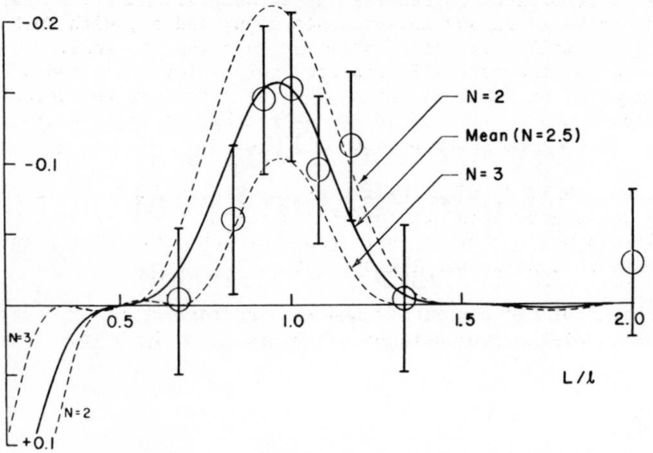

Fig. 1. Experimental results for the normalized correlation coefficient, together with theoretical curves for N = 2 and N = 3 and their mean.

In our experiment the transit time τ was approximately 3 nsec, while the photon rates r_1 and r_2 were of order 3 x 10^6 per sec. The 'exposure time' T was 20 μsec, which, in association with a photodetection efficiency of about 7%, resulted in about 10 detected photons per trial. It is of course important that this number should be significantly greater than unity if interference fringes are to be identified.

The light sources for the experiment were two single mode He:Ne lasers, whose beams were superposed at a small angle θ after passing through attenuators. An auxiliary arrangement involving an additional photodetector was used to measure the difference or beat frequency of the two beams. Whenever the beat frequency fell below 50 kc/sec the interference pattern was examined for 20 μsec. The interference detecting system consisted of a stack of thin glass plates, having their edges facing the beams and parallel to the interference fringes. They were arranged so that light falling on the odd-numbered plates was deflected to one photomultiplier, while light falling on the even-numbered plates was deflected to another. If the plate thickness $\frac{1}{2}L$ corresponds to a half-fringe width $\frac{1}{2}l$, and if the maxima of the interference fringes fall on the odd-numbered plates, say, then one phototube should register most of the photons and the other one practically none.

In fact the positions of the fringe maxima are not predictable and vary from trial to trial, so that the number of photons n_1 registered in one channel should decrease as the other number n_2 increases. Thus, if interference fringes are present, and if the fringe half spacing $\frac{1}{2}l$ corresponds to the plate thickness $\frac{1}{2}L$, there should be a negative correlation $\langle \Delta n_1 \Delta n_2 \rangle$. If $\frac{1}{2}l$ differs appreciably from $\frac{1}{2}L$, or if no interference fringes are formed, then $\langle \Delta n_1 \Delta n_2 \rangle$ should be close to zero, or perhaps positive.

Figure 1 shows the results of measurements of the correlation coefficients as a function of L/l, which was varied by changing the angle θ between the beams. It will be seen that a negative correlation is found when $L/l \approx 1$, which falls to zero once L/l departs appreciably from unity. Each experimental point is based on 400 measurements of n_1 and n_2, with $\langle n_1 \rangle$ and $\langle n_2 \rangle$ each about 5. Smaller values of $\langle n_1 \rangle$ and $\langle n_2 \rangle$ lead to smaller correlations. Also shown are the theoretically expected curves for N = 2 and N = 3, where N is the number of pairs of illuminated glass plates at the detector. In practice about 5 plates were actually covered by the light beams, so that the mean of the two curves is shown also.

It will be seen that, despite the rather low statistical accuracy of the results, there is good evidence for the existence of interference fringes, under conditions where there is negligible probability for the interaction of two or more photons within the apparatus.

A fuller account of the experiment and the underlying theory, together with some discussion of the implications of the result will be published elsewhere.

References

1. Magyar G and Mandel L 1963 *Nature 198* 255
2. Jordan T F and Ghielmetti F 1964 *Phys. Rev. Letters 12* 607
3. Paul H, Brunner W and Richter G 1963 *Ann. Physik 12* 325
4. Mandel L 1964 *Phys. Rev. 134* A10
5. Dirac P A M *Quantum Mechanics* (4th ed., Oxford University Press, London, 1958), chapter 1, p.9.
6. Mandel L and Wolf E 1965 *Rev. Mod. Phys. 37* 231

Reprinted from *Nature*, **198** (1963), 255–256

Interference Fringes Produced by Superposition of Two Independent Maser Light Beams

G. MAGYAR and L. MANDEL

Department of Physics, Imperial College of Science and Technology, London, UK

Optical interference effects are normally only observed with photons in coherent superposition states. Such states can be brought about with the help of beam splitters or induced optical transitions. It has been said that "each photon then interferes only with itself. Interference between different photons never occurs"[1]. However, transient interference effects have been demonstrated with two completely independent microwave beams[2]. Such effects, which are analogous to optical beats from incoherent sources[3-5], are immediately understandable in classical terms, for the amplitude and phase of each beam remains constant for a time short compared with the reciprocal frequency spread, $1/\Delta\nu$ (the coherence time). While it is true that in this case the ensemble average of the radiation intensity at different space points gives no indication of interference, the ensemble average is not relevant to any one short-time observation.

Interference effects ought to be observable also with two independent light beams, provided: (*a*) the photons in the two beams are not in orthogonal spin states; (*b*) the observation is made in a time shorter than the reciprocal total frequency spread, $1/\Delta\nu$, of the two beams, so that all the received photons may be regarded as falling into the same cell of phase space; (*c*) the mean number of photons received on a coherence area in a coherence time[6], that is, the mean occupation number of each cell of phase space - or the photon degeneracy parameter, δ - is much greater than 1.

As has been shown[7,8], for light from typical thermal sources δ is always much less than unity and reaches the value 1 only when the source temperature approaches 10^5 °K. On the other hand, δ is usually very large for optical maser beams[8], and this suggests the use of masers for the experiment[9]. We wish to report observation of interference fringes produced by the superposition of two independent beams of ruby maser light.

The experimental arrangement is shown in Fig. 1. Two light beams from two independent ruby masers are aligned with the help of two adjustable 45° mirrors and superposed on the photocathode of an electronically gated image tube[10]. The tube is magnetically focused and the image produced on the output fluorescent screen is photographed. The effective separation, d, of the two virtual (incoherent) sources is defined by two slits and the expected fringe spacing, x, is, as usual, given by

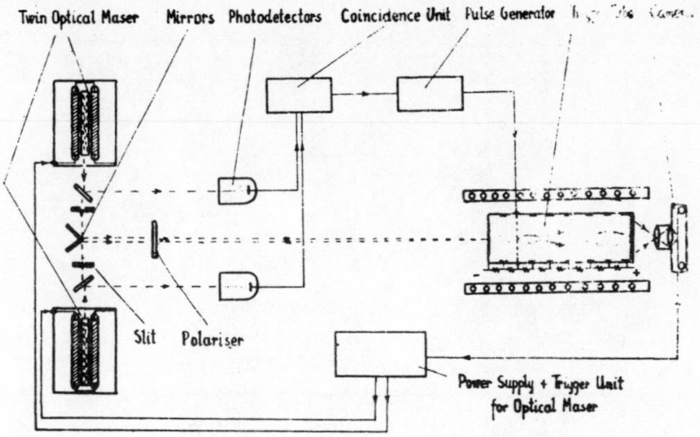

Fig. 1. Outline of the apparatus for recording transient interference fringes

$$x = \lambda D/d \qquad (1)$$

where λ is the wave-length and D the slit-to-cathode distance. The two rubies are rotated until the polarization planes of the two beams are parallel and a linear polarizer is introduced as a further precaution. A narrow-band optical filter attenuates most of the pumping radiation.

The flash tubes exciting the two masers are triggered simultaneously; but, as the maser emission from ruby is in the form of a series of random 'spikes' of approximately ½ μsec duration, two light beams are only occasionally emitted in coincidence. The image tube is therefore normally gated-off by a negative bias voltage applied to the grid. Two monitor photodetectors feeding into a coincidence circuit determine when two maser beams emerge in coincidence, and then cause the image tube to be gated-on by a positive pulse of selected duration (30-500 nsec) applied to the grid.

The two rubies used in the experiment are 5·7 cm long, of diameter 6 mm (0.02 per cent Cr, 60° orientation) and have silver-coated plane ends. They are surrounded by four straight xenon flash tubes in a configuration of the type described by Miles and Edgerton[11]. Under excitation of a few per cent above the threshold for maser action (approx. 200 joules) the rubies deliver about 20-40 'spikes', which leads to an average spike coincidence rate of about one per xenon flash. Provision is made for preventing double exposures resulting from double coincidences by paralysing the electronic gate for some hundreds of μsec after each exposure.

Fig. 2 shows a photograph of interference fringes obtained in a 40-nsec exposure, together with a micro-photometer tracing across the negative. The measured fringe spacing is 0.277 ± 0.003 mm, while the corresponding value calculated from equation (1) with D = 180 ± 0.5 cm, d = 4.51 ± 0.03 mm, λ = 6943 Å, is 0.277 ± 0.003 mm. The number of distinguishable fringes is about 23, which is rather less than the number expected from the ratio of slit spacing (4.51 mm) to slit width (about 0·3 mm). However, the high granularity

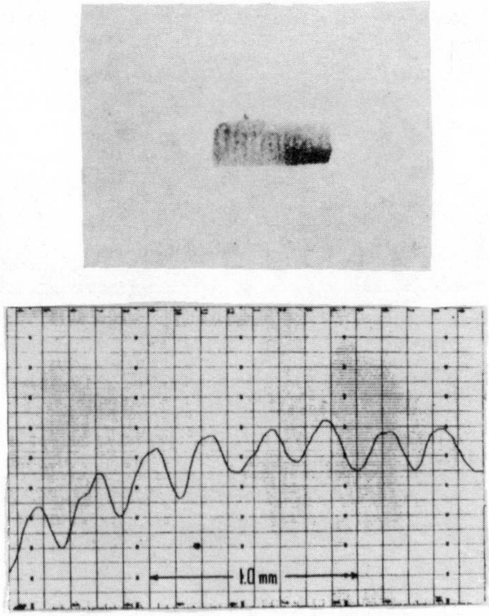

Fig. 2. An example of fringes recorded: (a) photograph; (b) microphotometer tracing.

of the film makes small variations of photographic density difficult to detect.

The maximum measured fringe visibility of the pattern is about 15 per cent. As in conventional interference experiments with coherent light beams, the visibility is partly determined by the ratio of the two light intensities, but in this case it is strongly dependent on the exposure time.

The approximate relation is easily calculated classically. Let us represent the two beams at some point in the superposition plane by the complex analytic random functions[12] $V_1(t - \frac{1}{2}\tau)$ and $V_2(t + \frac{1}{2}\tau)$, where $c\tau$ represents an optical path difference varying systematically from point to point. Then the resultant wave amplitude at any point of superposition is:

$$V(t) = V_1(t - \tfrac{1}{2}\tau) + V_2(t + \tfrac{1}{2}\tau) \tag{2}$$

and, if $I_1(t) = |V_1(t)|^2$ and $I_2(t) = |V_2(t)|^2$ are the instantaneous intensities, the resultant intensity is:

$$I(t) = I_1(t - \tfrac{1}{2}\tau) + I_2(t + \tfrac{1}{2}\tau) + 2R[V_1^*(t - \tfrac{1}{2}\tau)\, V_2(t + \tfrac{1}{2}\tau)] \tag{3}$$

Now $V_1(t)$ and $V_2(t)$ can be expressed in the form[12]:

$$\left.\begin{aligned} V_1(t) &= \sqrt{I_1(t)}\, \exp[2\pi i\nu_1 t + i\phi_1(t)] \\ V_2(t) &= \sqrt{I_2(t)}\, \exp[2\pi i\nu_2 t + i\phi_2(t)] \end{aligned}\right\} \tag{4}$$

where ν_1 and ν_2 are the mid-frequencies of the two light beams and $\phi_1(t)$ and $\phi_2(t)$ are random phase functions which, like $I_1(t)$ and $I_2(t)$, vary only slowly. In particular, they change by an insignificant amount in a time T short compared with the coherence time ξ. From (3) and (4), assuming that $\tau \ll \xi$, we have:

$$I(t) = I_1(t) + I_2(t) + 2\sqrt{[I_1(t)I_2(t)]}$$
$$\times \cos\,[2\pi(\nu_2 - \nu_1)t + \pi(\nu_2 + \nu_1)\tau + \phi_2(t) - \phi_1(t)] \qquad (5)$$

If the observation time T is also shorter than ξ, then the signal S(t,T) recorded in time T is given by:

$$\frac{S(t,T)}{T} = \frac{1}{T}\int_t^{t+T} I(t')dt'$$

$$= I_1(t) + I_2(t) + 2\sqrt{[I_1(t)I_2(t)]}\,\frac{\sin \pi(\nu_2 - \nu_1)T}{\pi(\nu_2 - \nu_1)T}$$
$$\cos[2\pi(\nu_2 - \nu_1)(t + \tfrac{1}{2}T) + \pi(\nu_2 + \nu_1)\tau + \phi_2(t) - \phi_1(t)] \qquad (6)$$

Since S(t,T)/T is a cosine function of τ, the recorded signal shows a positional modulation in the plane of superposition which we interpret as interference fringes. The fringe visibility is:

$$V = \frac{2}{\sqrt{[I_1(t)/I_2(t)]} + \sqrt{[I_2(t)/I_1(t)]}}\left|\frac{\sin \pi(\nu_2 - \nu_1)T}{\pi(\nu_2 - \nu_1)T}\right| \qquad (7)$$

and has a maximum value of unity when $I_1(t) = I_2(t)$ and $T \ll 1/|\nu_2 - \nu_1|$. Since $\phi_2(t)$ and $\phi_1(t)$ are random phases, the positions of the fringe maxima and minima are unpredictable.

This simple calculation ignores the possible presence of several axial modes in each maser beam. There is some evidence[13,14] that, close to threshold, the number of modes in a single maser 'spike' may be as low as unity, but if it is not the fringe visibility will be less than that given by (7).

Recent measurements[5,14-17] of the coherence time of the maser radiation from ruby have shown that this time is of the order of the spike duration, so that the conditions $T \ll \xi$ and $\tau \ll \xi$ under which equation (6) holds appear to have been satisfied in our experiment. But because the frequencies ν_2 and ν_1 refer to separate masers they may differ appreciably, and the requirement $T \ll 1/|\nu_2 - \nu_1|$ for high fringe visibility is normally the most difficult to satisfy. Not only are the optical lengths of the two rubies likely to differ, but also these lengths are subject to significant thermal variation[14]. At worst, $|\nu_2 - \nu_1|$ may be as great as the natural fluorescence line width, which would make γ extremely small. In practice, it is to be expected that thermal fluctuations will cause ν_2 and ν_1 to vary from pulse to pulse, and that observable interference fringes will only occasionally be obtained, and this is confirmed in our observations.

It seems, therefore, that, as with microwaves, these interference effects are describable in completely classical terms.

We thank Professor J.D. McGee and Dr. W.L. Wilcock for helpful suggestions. The work was supported financially by a grant from the Department of Scientific and Industrial Research.

REFERENCES

1. Dirac P A M, *Quantum Mechanics*, fourth ed. 9 (Clarendon Press, Oxford, 1958).
2. Hull G F 1949 *Amer. J. Phys. 17* 559
3. Forrester A T, Gudmundsen R A and Johnson P O 1955 *Phys. Rev. 99* 1961
4. Javan A, Ballik E A and Bond W L 1962 *Opt. Soc. Amer. 52* 96
5. Lipsett M S and Mandel L 1963 *Proc. Third Quantum Electronics Conf.*, Paris (to be published)
6. Mandel L and Wolf E 1962 *Proc. Phys. Soc. 80* 894
7. Gabor D *Progress in Optics 1* 111 (North Holland, Amsterdam, 1961)
8. Mandel L 1961 *J. Opt. Soc. Amer. 51* 797
9. Mandel L 1962 *J. Opt. Soc. Amer. 52* 1407
10. Mandel L 1961 *J. Soc. Motion Pict. and T.V.Eng. 70* 716
11. Miles P A and Edgerton H E 1961 *J. App. Phys. 32* 740
12. Born M and Wolf E *Principles of Optics 490* (Pergamon Press, London and New York, 1959)
13. Hughes T P 1962 *Nature 195* 325
14. Hanes G R and Stoicheff B P 1962 *Nature 195* 587
15. McMurtry B J and Siegman A E 1962 *App. Opt. 1* 51
16. Borie J C and Orszag A 1962 *C.R. Acad. Sci., Paris 255* 874
17. Berkley D A and Wolga G J 1962 *Phys. Rev. Letters 9* 479

Reprinted from *Phys. Rev.*, **D9** (1974), 853–860

Experimental Distinction between the Quantum and Classical Field Theoretic Predictions for the Photoelectric Effect*

J. F. CLAUSER

Department of Physics and Lawrence Berkeley Laboratory, University of California, Berkeley, CA 94720, USA

We have measured various coincidence rates between four photomultiplier tubes viewing cascade photons on opposite sides of dielectric beam splitters. This experimental configuration, we show, is sensitive to differences between the classical and quantum field-theoretic predictions for the photoelectric effect. The results, to a high degree of statistical accuracy, contradict the predictions by any classical or semiclassical theory in which the probability of photoemission is proportional to the classical intensity.

INTRODUCTION

It is commonly believed that experimental observations of the photoelectric effect establish the existence of uniquely quantum-mechanical properties of the electromagnetic field. Various classic experiments, coupled with the notion of microscopic energy conservation, are usually cited to establish this claim.[1] Unfortunately the insistence upon microscopic energy conservation amounts to an auxiliary criterion, which for a classical field theory (CFT) is inherently ambiguous. The quantum-mechanical energy of a photon, $h\nu$, is experimentally relevant to the photoelectric effect, determining the kinetic energy of the ejected electrons. This insistence, on the other hand, demands that the classical field energy $\int(E^2 + H^2)dV/8\pi$ be equal to $h\nu$ and be simultaneously conserved. The classical Maxwell equations contain no constraint that these energies be equal, as a quantum field theory (QFT) does.[2] This demand is, in fact, unreasonable for a classical field theory. It is therefore also unreasonable to use this constraint as a basis for an experimental distinction between the theories. With equal justification one might say that these experiments disprove microscopic energy conservation during the photoelectric process while upholding CFT. The above belief was finally shown to be totally unfounded when it was demonstrated that the above observations can be quantitatively accounted for by a semiclassical radiation theory

*Work supported by the U.S. Atomic Energy Commission.

in which the electromagnetic field is left unquantized.[3] The basic elements of this theory have since been used as a skeleton for the more recent and widely discussed neoclassical radiation theory (NCT) of Jaynes, Crisp, and Stroud.[4] In both of these theories it is hypothesized that the classical Maxwell equations describe the free electromagnetic field, and that this field never needs to be quantized to account for experimental observations. Previous experimental observations of the photoelectric effect, in and of themselves, are in agreement with this hypothesis, and do not appear to necessitate quantum-mechanical properties for the radiation field.

In 1955, following Schrödinger's suggestion, Ádám, Jánossy, and Varga[5] (AJV) searched for anomalous coincidences in a partially collimated beam of light.[5a] Jauch,[6] in his discussions of the foundations of quantum mechanics, has recently emphasized the importance of this experiment and an associated one performed by Jánossy and Náray[7] in establishing the existence of a wave-particle duality for photons. Moreover, the arguments of AJV and Jauch do not rely on energy conservation (although other assumptions are needed for their specific scheme) and as such are not subject to the above criticism. Attention is naturally called to this experiment by the recent discussions of semiclassical theories, in hopes that it might provide an additional aspect of the photoelectric effect upon which the predictions of CFT and QFT differ.

In this paper we will show that the actual values of the parameters for the arrangement of AJV (and subsequent similar experiments) unfortunately were insufficient to make that experiment conclusive. We then report new experimental results which are conclusive. Our measurements involved a comparison of various twofold coincidence rates between four photomultiplier tubes viewing cascade optical photons emitted by the same source through various beam splitters. We further show that this configuration is sensitive to differences between the QFT and CFT predictions for this effect without additional assumptions, such as those required by AJV. The results, to high statistical accuracy, contradict the predictions of any classical or semiclassical radiation theory in which the probability of photoemission is proportional to the classical field intensity. This includes, for example, NCT. Our experiment thus resurrects the photoelectric effect as a phenomenon requiring quantization of the electromagnetic field.

It is noteworthy that Aharonov et al.[8] presented a scheme similar to that of AJV as a *Gedanken-experiment*, while noting a paucity of actual experimental distinctions between CFT and QFT. The CFT prediction for our experiment follows reasoning similar to that by Titulaer and Glauber,[9] who discussed constraints applicable to CFT which demarcate a boundary between CFT and the more general QFT descriptions of the electromagnetic field.

In what follows we first contrast the CFT and QFT predictions for a single photon falling on a half-silvered mirror. We next discuss previous relevant experiments, contrast these with our own experimental scheme, and show that of these only ours provides the desired distinction. Finally we describe the apparatus and present the results.

PREDICTIONS FOR A SINGLE PHOTON FALLING ON A HALF-SILVERED MIRROR

In this section we review the arguments by AJV and Jauch. Consider the light emitted by a single atomic decay falling on a half-silvered mirror. During the decay a wave train (packet) of electromagnetic radiation is emitted. Suppose that it impinges upon a beam-splitting mirror, and that the two resultant wave trains are directed to two independent photomultipliers labelled γ_A and γ_B. We desire the QFT prediction for the $\gamma_A - \gamma_B$ coincidence rate. A simpler problem to consider first involves only the source atom and a second atom in one photocathode. We need the probability amplitude that, following deexcitation of the source atom, the second atom will become excited (or ionized). This has been obtained by Fermi[10] and Fano,[11] using the Wigner-Weisskopf approximation. The inclusion of a third atom in a second photocathode is then straightforward. Denote by S, A, and B, respectively, the ground states of the source atom and the two detector atoms, and by S^*, A^*, and B^* the corresponding excited or ionized states of these atoms. Initially the source atom is excited, and the two detector atoms are in their ground states; hence

$$|i\rangle = |S^*, A,B,0_1,\ldots,0_j,\ldots\rangle.$$

The remaining indices of the ket designate the state of the radiation field modes. The final state then has the form

$$\begin{aligned} |f\rangle = {} & U_A|S,A^*,B,0_1,\ldots,0_j,\ldots\rangle \\ & + U_B|S,A,B^*,0_1,\ldots,0_j,\ldots\rangle \\ & + U_S|S^*,A,B,0_1,\ldots,0_j,\ldots\rangle \\ & + \sum_j U_j|S,A,B,0_1,\ldots,1_j,\ldots\rangle. \end{aligned} \tag{1}$$

The various U_i can be evaluated from formulas found in Refs. 10 and 11. Thus QFT predicts that an observation will find at most one of the detector atoms ionized; i.e. coincident responses will occur only at the random accidental rate, induced by emissions from two different excited source atoms.[12]

Next we consider the same system from the CFT viewpoint. Our basic assumptions for this are twofold: (1) The electromagnetic field is described by the classical (unquantized) Maxwell equations, and (2) the probability of photoionization at a detector is proportional to the classical intensity of the incident radiation. These two assumptions alone are sufficient for our purposes, and they are in evident agreement with experiment.[13] Since ionizations at the γ_A and γ_B phototubes are independent, but are induced by nearly identical classical pulses of light, for a given split wave train both tubes will have roughly the same probability for registering a count. This independence implies that the probability that both will respond to the split wave train is simply the product of the probabilities that each will respond. The nonzero value of this product implies the existence of an anomalous coincidence rate above the accidental background. The CFT prediction is thus in marked contrast with QFT prediction, the latter requiring no coincidences above the background level.[12]

The above argument may be summarized very simply. Consider a radiation field quantum-mechanically with only one photon present. If we bring this into interaction with two separated atoms we will never get more than one photoelectron. If on the other hand we represent this field classically, we find that there is a nonvanishing probability for finding two photoelectrons. The classical Maxwell field has within it the possibility of providing with some probability any number of photons. Hence experiments of the above variety can distinquish between the two theories.

Such then is the argument of AJV and Jauch. Here we have also the basis for the usual particle interpretation of photons. A particle must be either transmitted or reflected. Both may be done simultaneously only by a wave. We then see how these macroscopic features of "particle-like" objects arise from the QFT formalism.

PREVIOUS EXPERIMENTAL RESULTS

That a photon is not split in two by a beam splitter is certainly "old hat", and it may seem surprising that we have gone to the effort to test this prediction experimentally. What is in fact much more surprising is that evidently no such experimental test has heretofore been performed, and such tests are clearly of great importance. Here we briefly review previous relevant experimental results and show that none provides the desired distinction.

Since the original work of AJV many two-photon coincidence experiments have been done, some involving light beams split by a half-silvered mirror. These all fall into two basic categories - atomic-cascade observations and Brown-Twiss-effect observations. Excellent reviews of these topics have been presented by Camhy-Val and Dumont[14] and Mandel and Wolf,[15] respectively. Cascade-photon observations in their usual configuration are not suitable for the above test, since in these, two different unsplit photons are observed.

The AJV experiment, although intended as a test of the above scheme, actually served as a fore-runner to the Brown-Twiss-effect experiments. Figure 1 reproduces a diagram of the experiment of AJV. In it they selected the light of a single spectral line with a monochromator, and focussed it through a beam splitter onto two photomultipliers whose outputs drove a coincidence circuit.

Let us evaluate the magnitude of the expected anomalous coincidence rate. The CFT predictions for one and two photodetectors sharing the same field were discussed earlier by Mandel[16] from the above fundamental assumptions. Denote by I(t) the instantaneous classical intensity incident simultaneously upon the γ_A and γ_B detectors due to their illumination by the whole source volume. The singles rates for the A and B detectors, averaged over their response time T, is given by

$$S_A = \alpha_A T^{-1} \int_{-T/2}^{T/2} \langle I(t+t')\rangle dt',$$

$$S_B = \gamma_B T^{-1} \int_{-T/2}^{T/2} \langle I(t+t')\rangle dt', \tag{2}$$

where α_A and α_B are measures of the detector efficiencies, and the angular brackets denote an ensemble average over the emitted intensities. Similarly, the average coincidence rate as a function of event separation τ is given by

$$C_{AB}(\tau) = \alpha_A \alpha_B T^{-1} \int_{-T/2}^{T/2} \int_{-T/2}^{T/2} \langle I(t+t')I(t+t''+\tau)\rangle dt'dt''. \tag{3}$$

To obtain a model-independent prediction for the coincidence rate only from data on the singles rates does not appear possible, since (2) and (3) involve different averages of I(t). AJV thus had to make various assumptions (assump-

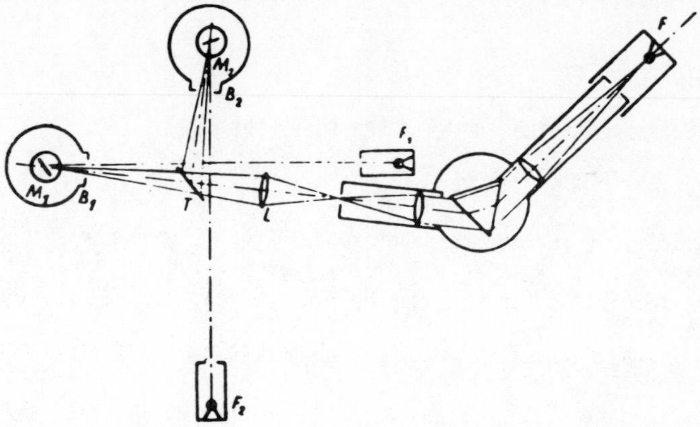

Fig.1. Experimental arrangement of Ádám, Jánossy, and Varga. Light from source F is focussed through a monochromator onto photomultipliers M_1 and M_2 via beam splitter T. (Figure after Ádám, Jánossy, and Varga.)

tions which were unnecessary in the case of our own experiment). They tacitly assumed that

$$\left(\int_{-T/2}^{T/2} \langle I(t+t')\rangle dt'\right)^2 \approx \int_{-T/2}^{T/2}\int_{-T/2}^{T/2} \langle (t+t')I(t+t''+\tau)\rangle dt'dt'' \tag{4}$$

holds for each decay, when τ is the order of the decaying state lifetime. If then E pulses per second are emitted per unit time by a source, and if η is the average probability that a photomultiplier will yield a count, given an atomic decay, the count rate at that detector is

$$S = E\eta. \tag{5}$$

The expected anomalous coincidence rate predicted by the AVJ assumptions is then given approximately by

$$C \approx \eta^2 E. \tag{6}$$

Assuming negligible detector dark rates, the accidental coincidence background rate from which C must be distinguished is

$$A \approx \eta^2 E^2 2\tau_c, \tag{7}$$

where τ_c is the resolving time of the coincidence system. One can now calculate the integration time required to measure to a precision of N standard deviations the difference between the excess coincidence rate given by Eq.(6) and the zero excess rate predicted by QFT. Doing this we obtain

$$T_{int} \approx (1 + 4E\tau_c)N^2\eta^{-2}E^{-1}, \tag{8}$$

which in the limit of high source rates takes the form

$$T_{int} \approx 4N^2\tau_c/\eta^2. \tag{9}$$

Thus the validity of their experiment rests directly upon the assumed or measured value of η: If it is too small, T_{int} will be too long and the experiment will see only the random accidental background. AJV measured their detector efficiencies by assuming that these were given by the formula

$$\eta = Rh\nu/W, \tag{10}$$

where R is the count rate obtained for a given beam of photons, and W is the power in the same beam measured bolometrically. They thus found $\eta = 1/300$. With a resolving time $\tau_c = 2.3$ μsec one calculates $T_{int} = 20.7$ sec for $N = 5$. From this reasoning AJV felt confident that they should have observed the anomalous coincidence rate, if it was present.

Let us reexamine from the CFT viewpoint the assumption tacitly contained in Eq.(10). Although the introductory arguments did not contain a requirement for energy conservation, AJV have unnecessarily reintroduced it with this assumption; this is in direct conflict with our fundamental assumptions for a CFT. In our derivation above, η is the probability for a detector response, given a source atom decay. Clearly a wavelike pulse emitted by a source atom will expand, in the worst case spherically, or at best with a radiation pattern having a preferred direction.[17] Much of this pulse will not enter the narrow acceptance solid angle subtended by the monochromator. Propagation will cause it to suffer an enormous decrease in intensity, commensurate with its expansion. Assuming macroscopic energy conservation on the average, the power W should then represent the total average power radiated by the source at the appropriate wavelength, not that which happens to be measured within the beam itself. The number calculated from Eq.(10) must be appropriately decreased by the fraction of the solid angle effectively subtended by the detectors. Other optical losses will decrease this number even further.

If we conservatively estimate from their diagram the solid-angle loss to be 1/400, their actual detector efficiency for spherically emitted wavelike pulses was undoubtedly less than 8×10^{-6} in which case the required integration time for even $N = 1$ becomes $T_{int} \approx 1.3 \times 10^5$ sec. This is an order of magnitude longer than the duration of their experiment. Thus the experiment of Ádám, Jánossy, and Varga appears to be inconclusive when reexamined in this light.

A similar analysis applies to the experiments of Givens and of Brannen and Ferguson.[5a] In the x-ray coincidence experiment of Givens, the source solid angle viewed by the detector pair was $\approx 3.5 \times 10^{-5}$ sr, smaller than that of AJV. Combining this with his $\approx 15\%$ quantum efficiencies, we estimate the overall detector efficiencies to be $\approx 2.1 \times 10^{-7}$ (neglecting the appreciable loss due to the beam-splitting crystal). Givens employed a resolving time of $\approx 1.7 \times 10^{-4}$ sec. From Eq.(9), we find then that an integration time of nearly 500 yr is required for this apparatus to produce results with a confidence level corresponding to just one standard deviation. Similar reasoning finds the actual integration time of Brannen and Ferguson deficient by a factor $\approx 1.7 \times 10^5$. These experiments are thus likewise inconclusive for deciding the above question.

Finally let us consider experiments of the Brown-Twiss variety. These experiments have a configuration basically the same as that of AJV. Because of the nature of this effect, however, all existing data have been accumulated with detectors subtending extremely small solid angles, much smaller even than those of AJV. From Eq.(9) we see that the required integration time scales with the inverse square of the detector solid angles; hence it would be hopeless to try to search for the above anomalous coincidence rate with such arrangements. Furthermore, in these experiments, the Brown-Twiss effect itself would tend to mask the effect we seek. In summary, then, none of the above experiments can provide the desired distinction.

EXPERIMENTAL SCHEME REQUIRING NO ADDITIONAL ASSUMPTIONS

The above discussion indicates that an observation of the anomalous coincidences predicted by a CFT requires highly efficient photodetectors. However, even if AJV had had the required efficiency and integration time, their experimental arrangement necessitated assumptions concerning the various field averages, and hence assumed a basic model for the emission mechanism. Since no universally acceptable model is at hand, we have chosen to employ a scheme which renders our results model-independent. We did this by "splitting" simultaneously both the first and second photons of an atomic cascade. We viewed the light emitted on opposite sides of an assembly of excited atoms and focussed it separately into two beams. The wavelength λ_1 on one side was selected to correspond to that of the first transition of the cascade, and that on the other, λ_2, to the second. The two light beams impinged on beam splitters, thus creating a total of four beams. Four associated photomultipliers labelled γ_{1A}, γ_{1B}, γ_{2A}, and γ_{2B} detected them. We monitored the coincidence rates between the four combinations: $\gamma_{1A}-\gamma_{1B}$, γ_{2A}-γ_{2B}, γ_{1A}-γ_{2B}, and γ_{2A}-γ_{1B}. A diagram of the arrangement is shown in Fig.2.

Define $I_1(t)$ and $I_2(t)$ as the instantaneous intensity at the γ_{1A}-γ_{1B} beam splitter with wavelength λ_1, and that at the γ_{2A}-γ_{2B} beam splitter with wavelength λ_2, respectively. It follows directly from the Cauchy-Schwarz inequality that the following inequality holds:

$$\left[\int_{-T/2}^{T/2}\int_{-T/2}^{T/2} \langle I_1(t+t'+\tau_1)I_1(t+t''+\tau_1)\rangle dt'dt''\right]$$

$$\left[\int_{-T/2}^{T/2}\int_{-T/2}^{T/2} \langle I_2(t+t'+\tau_2)I_2(t+t''+\tau_2)\rangle dt'dt''\right]$$

$$\geq \left[\int_{-T/2}^{T/2}\int_{-T/2}^{T/2} I_1(t+t'+\tau_1)I_2(t+t''+\tau_2)\rangle dt'dt''\right]^2 \tag{11}$$

Using (3), we can write this as

$$C_{1A-1B}(0)C_{2A-2B}(0) \geq C_{1A-2B}(\tau)C_{1B-2A}(\tau). \tag{12}$$

Here we have ignored a possible polarization dependence of the detectors, and the finite photocathode areas, as well as the nonvanishing phototube dark rates. It can be shown that the inequality (12) may be summed over these contributions without change of form. Thus it is fully general and holds for these cases as well. The coincidence rates C_{1A-2B} and C_{2A-1B} here are the nonvanishing cascade rates. The product of these sets a lower bound to the product of the anomalous rates C_{1A-1B} and C_{2A-2B}. Thus, CFT predicts a large anomalous coincidence rate satisfying (12). The prediction of QFT significantly violates this inequality, requiring no coincidences except those due to two-atom excitations.

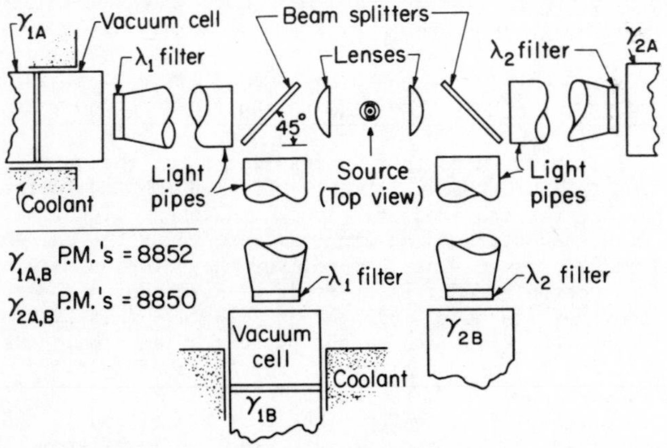

Fig.2 Schematic diagram of our apparatus.

APPARATUS AND RESULTS

Figure 2 is a diagram of the apparatus. The source contained ^{202}Hg atoms which were excited by electron bombardment. Light produced at $\lambda_1 = 5676$ Å and $\lambda_2 = 4358$ Å by the cascade 9 $^1P_1$ - 7^3S_1 - 6^3P_1 was used. It was made parallel by lenses (aspheric, $f \approx 1$), and fell on TiO_2-coated glass beam splitters (transmission ≈ 63% and 35% for opposite linear polarizations, inclined at 45° to the incident beams). Each resulting beam was directed through an interference filter [transmission ≈ 50% at 5676 Å, full width at half maximum (FWHM) ≈ 50 Å for γ_{1A} and γ_{1B}; transmission ≈ 30% at 4358 Å, FWHM ≈ 100 Å for γ_{2A} and γ_{2B}] onto an appropriate photomultiplier tube [RCA 8852, quantum efficiency (QE) ≈ 15% at 5676 Å, dark current ≈ 50-300 Hz, operated at -80 °C for γ_{1A} and γ_{1B}; RCA 8850, QE ≈ 30% at 4358 Å, dark current ≈ 100 Hz, operated at 20 °C for γ_{2A} and γ_{2B}].

The source itself was patterned after a design by Holt, Nussbaum, and Pipkin,[18] and was made by using standard techniques. The electron gun was a standard 10-W cathode-ray tube gun obtained through the courtesy of the Raytheon Corporation. It was mounted with suitable deflecting electrodes and light masks in a quartz and Pyrex envelope, evacuated, and cleaned by baking and discharging; the metal parts were outgassed by induction heating, and the oxide cathode was activated. A few milligrams of 93%-pure ^{202}Hg were then distilled into the tube and the envelope sealed. The Hg vapour pressure was controlled by keeping a side arm immersed in ice water. A beam current of approximately 0.7μA traversed the cylindrical excitation region (length ≈ 2 mm, diam. ≈ 1 mm). The light output was stable. Photomultipliers operating in coincidence were separated from each other by more than 1.5 m to eliminate anomalous coincidences caused by cosmic rays. Light pipes minimized the light loss during transit. The interference filters were placed at the outer ends of the light pipes to minimize anomalous coincidences due to scintillations in the beam splitter and collimator lenses. These could be excited by cosmic rays and/or residual radioactivity therein. This configuration also effectively eliminated phototube cross talk induced by light emitted at the last dynodes. High-speed electronics with ≈ 1-nsec resolving time were used. The discriminators drove a time-to-amplitude converter whose output was fed to a pulse-height analyzer. External slow coincidence circuits gated the signals into one of the four analyzer memory quadrants, corresponding to the particular coincidence mode. The analyzer thus simultaneously accumulated the four different delayed coincidence spectra, i.e. the number of events pairs as a function of event separation time.

The results, shown in Fig.3(a)-3(d), represent more than 26 hours of integration. We find no evidence for an anomalous coincidence rate in either the γ_{1A}-γ_{1B} or the γ_{2A}-γ_{2B} mode, but the normal cascade mode is quite apparent. For a timing and sensitivity check, both tube pairs were excited through the beam splitters by short-duration "classical" light pulses from a barium-titanate source,[19] with approximately one photon per pulse. The resultant coincidence spectra are shown in Figs. 3(e) and 3(f). Finally, Fig.3(g) shows that our data severely violate (12) for a wide range of delays τ.

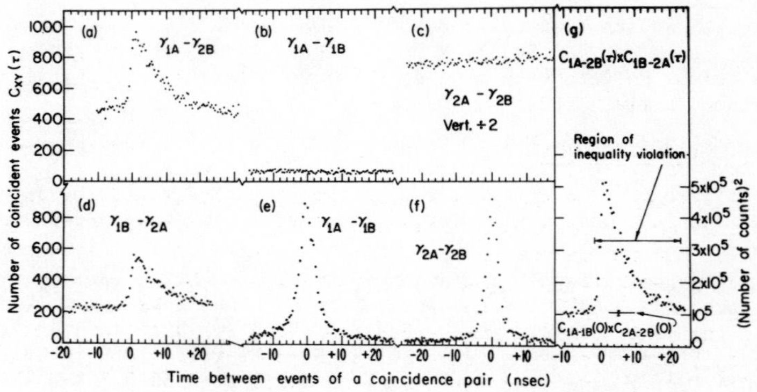

Fig.3. (a)-(d) Time-delay coincidence spectra of the four monitored channels: C_{1A-2B}, C_{1A-1B}, C_{2A-2B}, and C_{1B-2A}. (e)-(f) C_{1A-1B} and C_{2A-2B} coincidence spectra in response to short pulses of light incident upon beam splitters produced by a barium titanate source. (g) Product of C_{1A-2B} and C_{1B-2A} versus time delay. For small times this clearly exceeds the indicated value of the product C_{2A-2B} and C_{1A-1B} evaluated at zero delay.

DISCUSSION

The importance of experimentally demonstrating phenomena which require a quantization of the electromagnetic field has been emphasized recently by a number of suggestions that such a quantization is unnecessary. Many standard effects have thus been challenged as not providing definitive proof for the necessity of this quantization.[4,20] Several recent experiments testing the specific predictions of NCT and the Schrödinger interpretation have been performed[21] in this direction. The present experiment and others[20] have tested the quantum-mechanical aspects of Maxwell's equations. So far, none has uncovered any departure from the quantum-electrodynamic predictions, but severe departures from CFT predictions have been found. The classical (unquantized) Maxwell equations thus appear to have only limited validity.

ACKNOWLEDGEMENTS

The author thanks J.A. Crawford and M.H. Prior for helpful and stimulating discussions during the performance of this experiment.

REFERENCES

1. See, for example, Davis W and Mandel L in *Coherence and Quantum Optics: Proceedings of the 3rd Rochester Conference on Coherence and Quantum Optics,* edited by Mandel L and Wolf E (Plenum, New York, 1973), p.113 and references therein.

2. See, for example, Heitler W, *Quantum Theory of Radiation,* 3rd edition (Oxford, London, 1954), p.58.

3. Mandel L, Sudarshan E C G and Wolf E 1964 *Proc. Phys. Soc. Lond. 84* 435; Lamb W E and Scully M O in *Polarization: Matiere et Rayonnement,* edited by Société Francaise de Physique (Presses Universitaires de France, Paris, 1969).

4. Crisp M D and Jaynes E T 1969 *Phys. Rev. 179* 1253; 1969 *Phys. Rev. 185* 2046; Stroud C R Jr. and Jaynes E T 1970 *Phys. Rev. A 1* 106; Jaynes E T; 1970 *ibid. 2* 260. See also the extensive review by Jaynes E T in *Coherence and Quantum Optics* (see Ref.1), p.35, and references therein.

5. Ádám A, Jánossy L and Varga P 1955 *Acta Phys. Hung. 4* 301; 1955 *Ann. Physik 16* 408.

5a. Similar experiments were also performed by Brannen E and Ferguson H I S [1956 Nature (Lond.) *178* 481], and earlier with x rays by Givens M P [1946 Philos. Mag. *37* 335].

6. Jauch J M in *Foundations of Quantum Mechanics, International School of Physics "Enrico Fermi",* edited by d'Espagnat B (Academic, New York, 1971), p.20; Jauch J M *Are Quanta Real*? (Indiana Univ. Press, Bloomington, Indiana, 1973).

7. Jánossy L and Náray Zs. 1957 *Acta Phys. Hung. 7* 403.

8. Aharonov Y, Falkoff D, Lerner E and Pendleton H 1966 *Ann. Phys. (N.Y.) 39* 498.

9. Titulaer U M and Glauber R J 1965 *Phys. Rev. 140* B676.

10. Fermi E 1932 *Rev. Mod. Phys. 4* 87.

11. Fano U 1961 *Am. J. Phys.* *29* 539

12. An additional coincidence rate, the Brown-Twiss effect, is also due to two-atom excitations (see Ref.11). Its contribution to large-aperture-large-solid-angle systems such as those under discussion here is negligible.

13. Forester A T Gudmundsen R A and Johnson P O 1955 *Phys. Rev.* *99* 1691.

14. Camhy-Val C and Dumont A M 1970 *Astron. and Astrophys.* *6* 27.

15. Mandel L and Wolf E 1965 *Rev. Mod. Phys.* *37* 231. See also *Selected Papers on Coherence and Fluctuations of Light*, edited by Mandel L and Wolf E (Dover, New York, 1970), Vols. I and II.

16. Mandel L 1958 *Proc. Phys. Soc. Lond.* *72* 1037. More recent discussions are summarized by Klauder J R and Sudarshan E C G *Fundamentals of Quantum Optics* (Benjamin, New York, 1968), Chap.3.

17. One might assume a model in which all decays are identical, with the emitted intensity nearly isotropic. On the other hand, recent direct observations of atomic recoil associated with spontaneous emission indicate that, in a semiclassical scheme, either the radiation must be emitted with a preferred direction, or this recoil is due to some other mechanism [Picque J L and Vialle J L 1972 *Opt. Commun.* *5* 402; Schieder R, Walther H and Wöste L 1972 *ibid.* *5* 337; Frisch R 1933 *Z. Physik* *86* 42]. These experiments were in response to an elegant proof by Einstein A [1917 *Phys. Z.* *18* 121] showing that recoil is necessary if thermal equilibrium is to be maintained when a dilute gas interacts with radiation. (His arguments also relied upon microscopic energy conservation.) Various schemes to account for such beaming have been proposed. See Oseen C W 1932 *Ann. Physik* *69* 202 and the more recent work by Beers Y 1972 *Am. J. Phys.* *40* 1139; 1973 ibid *41* 275.

18. Holt R A Ph.D. thesis, Harvard University, 1973 (unpublished); Nussbaum G H and Pipkin F M 1967 *Phys. Rev. Lett.* *19* 1089.

19. Innes T G and Cox G C in Lawrence Berkeley Laboratory Counting Handbook, LBL Report No. UCRL-3307 Rev. (unpublished), pp. cc 8-31

20. Clauser J F 1972 *Phys. Rev. A* *6* 49.

21. Wessner J M, Anderson D K and Robiscoe R T 1972 *Phys. Rev. Lett.* *29* 1126; Gibbs 1972 *ibid.* *29* 459, and in *Coherence and Quantum Optics* (see Ref. 1), p.83; Gibbs H M, Churchill G G and Salamo G J 1973 *Phys. Rev. A* *7* 1766; Gaviola E 1928 *Nature* *12* 772. This last experiment is discussed by Wessner J M, Anderson D K and Robiscoe R T in 1973 *Phys. Today* *26* (No.2), 13.

Reprinted from *Proc. Phys. Soc.*, **84** (1964), 435–444

Theory of Photoelectric Detection of Light Fluctuations*

L. MANDEL**, E. C. G. SUDARSHAN*** and E. WOLF****

**Department of Physics, Imperial College, London, UK

***Department of Physics, Brandeis University, Waltham, Massachusetts, USA

****Department of Physics and Astronomy, University of Rochester, Rochester, New York, USA

The basic formulae governing the fluctations of counts registered by photoelectric detectors in an optical field are derived. The treatment, which has its origin in Purcell's explanation of the Hanbury-Brown Twiss effect, is shown to apply to any quasi-monochromatic light, whether stationary or not, and whether of thermal origin or not. The representation of the classical wave amplitude of the light by Gabor's complex analytic signal appears naturally in this treatment.

It is shown that the correlation of counts registered by N separate photodetectors at N points in space is determined by a 2Nth order correlation function of the complex classical field. The variance of the individual counts is shown to be expressible as the sum of terms representing the effects of classical particles and classical waves, in analogy to a well-known result of Einstein relating to black-body radiation. Since the theory applies to correlation effects obtained with any type of light it applies, in particular, to the output of an optical mase, although, for a maser operating on one mode, correlation effects are likely to be very small.

1. INTRODUCTION

In an important note Purcell (1956) has given a very clear explanation of an effect first observed by Brown and Twiss (1956), namely the appearance of correlation in the fluctuations of two photoelectric currents evoked by coherent beams of light. The method employed by Purcell is semi-classical, but it brings out the essence of the phenomenon much more clearly than most other approaches. Purcell's method has been developed further by Mandel (1958, 1959, 1963a,b) and applied to the analysis of related problems by Alkemade (1959) and by Wolf (1960) (see also Kahn 1958).

---

*This research was supported in part by the U.S. Army Research Office (Durham).

**Now at the Department of Physics and Astronomy, University of Rochester, Rochester, New York, U.S.A.

In all the publications just referred to, the light incident on the photodetector was implicitly or explicitly assumed to be of thermal origin. However, in view of the development of optical masers the question has been raised in recent months as to the existence of the Hanbury Brown-Twiss effect with light from other sources. In this connection Glauber (1963a,b) has expressed doubts about the applicability of these stochastic semi-classical methods to the analysis of fluctuation and correlation experiments obtained with light from non-thermal sources. Although partial answers to Glauber's critical remarks have already been given (Mandel and Wolf 1963, Sudarshan 1963 a,b; see also Jaynes and Cummings 1963), it is, of course, desirable to examine this semi-classical approach more closely. This is done in the present paper where new results on fluctuations and correlations are also obtained. The main conclusions are:

(i) In the conditions under which light fluctuations are usually measured by photoelectric detectors, the semi-classical treatment applies as readily to light of non-thermal origin as to thermal light, and to non-stationary as well as to stationary fields.

(ii) The representation of the optical field by complex analytic signals (as customary in the classical theory of optical coherence) appears naturally in this treatment.

(iii) There exists a very simple formula for the variance of the number of photoelectrons registered by a photodetector in a given time interval, when it is illuminated by any quasi-monochromatic light beam. This formula expresses the variance as the sum of two terms, one of which can be interpreted as representing the effect of fluctuations in a system of classical particles, and the other as arising from the interference of classical waves. This result, which may also be interpreted as describing the fluctuations of the light itself, is strictly analogous to a well-known result first established by Einstein (1909 a,b) for energy fluctuations in an enclosure containing black-body radiation, under conditions of thermal equilibrium.

(iv) The correlation between the number of photoelectrons registered by N separate photodetectors is expressible in terms of functions of 2Nth order correlations of the classical complex wave amplitudes.

The simplicity of the semi-classical theory and its wide range of validity makes it well suited for the analysis of many problems relating to photoelectric detection of light fluctuations. Our work demonstrates that a full quantum field theoretical treatment is not at all necessary for the analysis of such problems.

2. THE PROBABILITY DISTRIBUTION FOR PHOTOELECTRONS EJECTED FROM A PHOTO CATHODE ILLUMINATED BY A LIGHT BEAM

First, consider an electromagnetic wave interacting with a quantum mechanical system, playing the role of a 'detector', in a bound state $|\psi_b\rangle$. Suppose that $|\psi_b\rangle$, together with a continuum of unbound states $|\psi_\kappa\rangle$, form a complete set of orthonormal eigenstates of the unperturbed time-independent Hamiltonian H_0 of the system, i.e.

$$H_0|\psi_\kappa\rangle = E_\kappa|\psi_\kappa\rangle. \tag{2.1}$$

Here κ stands collectively for all the indices labelling the eigenstates of

H_0. If a time-dependent perturbation $H_1(t)$ is applied to the system at time t_0, then the state at time t may be expressed in terms of the set $|\psi_\kappa\rangle$:

$$|\psi(t)\rangle = \int C_\kappa(t)\exp\left(-\frac{iE_\kappa t}{\hbar}\right)|\psi_\kappa\rangle\, d\kappa + C_b(t)\exp\left(-\frac{iE_b t}{\hbar}\right)|\psi_b\rangle. \quad (2.2)$$

The integration with respect to κ is to be interpreted as an integration over ω_k, where

$$\hbar\omega_k = E_\kappa - E_b \geq 0, \qquad E_\kappa \equiv E_{k\mu}$$

and a summation over μ, where μ denotes the set of all quantum numbers other than k. The coefficients $C_\kappa(t)$ are given by the familiar formula of first-order perturbation theory

$$C_\kappa(t) = \frac{1}{i\hbar}\int_{t_0}^{t} \langle\psi_\kappa|H_1(t')|\psi_b\rangle \exp(i\omega_k t')dt'. \quad (2.3)$$

Let $\rho(\omega_k)d\omega_k$ be the number of states $|\psi_\kappa\rangle$ in the energy interval $\hbar\, d\omega_k$. Then the probability that a transition has occurred to any of the unbound states by the time t is

$$\int_0^\infty |C_\kappa(t)|^2\rho(\omega_k)d\kappa.$$

If $t - t_0 = \Delta t$ we define a transition probability per unit time by

$$\Pi(t) = \frac{1}{\Delta t}\int_0^\infty |C_\kappa(t)|^2\rho(\omega_k)\, d\kappa \quad (2.4)$$

Both $C_\kappa(t)$ and $\Pi(t)$ depend also on the position of the atomic nucleus (to be specified by position vector $\underline{R}$ later on).

Let us now consider the interaction between a *single* atom and an incident electromagnetic wave, represented by the vector potential $\underline{A}(\underline{r},t)$. Let the momentum of a typical electron of the atom be represented by $\underline{p}$. Then the interaction Hamiltonian, in the usual notation is

$$H_1(t) = \frac{e}{mc}\ \underline{A}(\underline{r},t).\underline{p}. \quad (2.5)$$

We now express $\underline{A}(\underline{r},t)$ in the form of a Fourier integral

$$\underline{A}(\underline{r},t) = \int_{-\infty}^{\infty} A\,(\underline{r},\omega)e^{i\omega t}d\omega. \quad (2.6)$$

Since $\underline{A}$ is real,

$$A\,(\underline{r},-\omega) = A^*(\underline{r},\omega). \quad (2.7)$$

Then, from (2.3), (2.5) and (2.6), we obtain

$$C_\kappa(t) = \frac{e}{i\hbar mc}\int_{t_0}^{t_0+\Delta t} dt' \int_{-\infty}^{\infty} d\omega \exp\{i(\omega_k-\omega)t'\} M_\mu(\underline{R},\omega,\omega_k)$$

$$= \frac{e\Delta t}{i\hbar mc}\int_{-\infty}^{\infty} d\omega \exp\{i(\omega_k-\omega)(t_0+\tfrac{1}{2}\Delta t)\} \frac{\sin\{\frac{1}{2}(\omega_k-\omega)\Delta t\}}{\frac{1}{2}(\omega_k-\omega)\Delta t.} M_\mu(\underline{R},\omega,\omega_k) \tag{2.8}$$

where

$$M_\mu(\underline{R},\omega,\omega_k) = \langle\psi_\kappa | A\ (r,\omega).\underline{p}|\psi_b\rangle \tag{2.9}$$

is the matrix element between the ground state and the continuum state and is a function of $\underline{R}$, the position of the atomic nucleus. Now provided $\Delta t \gg 1/\omega_k$ for all k for which $|M_\mu(\underline{R},\omega,\omega_k)|^2$ is appreciable, the integrand considered as a function of ω_k will be very sharply peaked about ω. In practice the smallest value of ω_k likely to matter when we are dealing with photoelectric transitions is of the order of 10^{14} c/s. The condition $\omega_k\Delta t \gg 1$ is therefore likely to hold for all measurable intervals Δt. The contribution to $C_\kappa(t)$ arises from values of ω in the neighbourhood of ω_k, and since $\omega_k > 0$ we may replace the lower limit in (2.8) by zero. From (2.4) and (2.8),

$$\Pi(t) = \frac{e^2\Delta t}{\hbar^2m^2c^2}\int_o^\infty\int_o^\infty\int_o^\infty d\omega d\omega' d\omega_k \exp\{i(\omega'-\omega)(t_0+\tfrac{1}{2}\Delta t)\frac{\sin\{\frac{1}{2}(\omega_k-\omega)\Delta t\}}{\frac{1}{2}(\omega_k-\omega)\Delta t}$$

$$\times \frac{\sin\{\frac{1}{2}(\omega_k-\omega')\Delta t\}}{\frac{1}{2}(\omega_k-\omega')\Delta t}\sum_\mu \rho(\omega_k) M^*_\mu\ (\underline{R},\omega',\omega_k) M_\mu(\underline{R},\omega,\omega_k). \tag{2.10}$$

It is evident that the integrand effectively vanishes unless ω, ω' and ω_k are nearly equal, to within an amount of the order of $1/\Delta t$. It is reasonable to suppose that $\rho(\omega_k)$ does not vary significantly over such a small range of ω_k and may therefore be replaced by $\{\rho(\omega)\rho(\omega')\}^{\frac{1}{2}}$ in (2.10).

Next let us assume that the incident radiation is quasi-monochromatic, i.e. its effective bandwidth is small compared with the mid-frequency. Then we choose $1/\Delta t$ small compared with any frequency ω that contributes to (2.10), but large compared with the difference $\omega - \omega'$ between any pair of frequencies ω and ω', and (2.10) then reduces to

$$\Pi(t) = \frac{2\pi e^2}{\hbar^2m^2c^2}\int_o^\infty\int_o^\infty d\omega d\omega' \exp\{i(\omega'-\omega)t\}\{\rho(\omega)\}^{\frac{1}{2}}\{\rho(\omega')\}^{\frac{1}{2}}\frac{\sin\{\frac{1}{2}(\omega-\omega')\Delta t}{\frac{1}{2}(\omega-\omega')\Delta t}$$

$$\times \sum_\mu\ M^*_\mu\ (\underline{R},\omega',\omega') M_\mu(\underline{R},\omega,\omega). \tag{2.11}$$

Now we have the identity

$$\frac{\sin\{\frac{1}{2}(\omega-\omega')\Delta t\}}{\frac{1}{2}(\omega-\omega')\Delta t} = \frac{1}{\Delta t}\int_{-\Delta t/2}^{\Delta t/2} \exp\{i(\omega'-\omega)\tau\}d\tau.$$

Hence the probability of a photoelectric transition per unit time may be expressed in the form

$$\Pi(t) = \frac{2\pi e^2}{\hbar^2 m^2 c^2} \sum_{\mu} \frac{1}{\Delta t} \int_{-\Delta t/2}^{\Delta t/2} W_{\mu}{}^*(\underline{R}, t+\tau) W_{\mu}(\underline{R}, t+\tau) d\tau \tag{2.12a}$$

where

$$W_{\mu}(\underline{R}, t) = \int_{0}^{\infty} d\omega \; e^{-i\omega t} \{\rho(\omega)\}^{\frac{1}{2}} M_{\mu}(\underline{R}, \omega, \omega). \tag{2.12b}$$

Since the Fourier spectrum of $W_{\mu}(\underline{R},t)$ contains no negative frequencies, W_{μ} is, according to a well-known theorem (Titchmarsh 1948), analytic and regular in the lower half of the complex t plane.

So far we have been considering the interaction between an incident electromagnetic field and a single atom only. Now suppose that we are dealing with a plane wave incident normally on an extended detector, which is in the form of a thin photoelectric layer containing a large number of electrons in initial states $|\psi_b\rangle$. With the assumption that these atoms may be treated as independent (i.e. that their electron wave functions do not appreciably overlap) and that the states are not appreciably depopulated, we can express the probability $P(t)\Delta t$ of photoelectric detection from any part of the photosurface by

$$P(t) = \frac{2\pi e^2 N}{\hbar^2 m^2 c^2 \Delta t} \sum_{\mu} \int_{-\Delta t/2}^{\Delta t/2} W_{\mu}^* (\underline{R}, t+\tau) W_{\mu}(\underline{R}, t+\tau) d\tau \tag{2.13}$$

where N is the number of effective electrons. Since we are dealing with an incident plane wave, the right-hand side of (2.13) is independent of $\underline{R}$.

In the dipole approximation we may express the matrix element M_{μ}, given by (2.9), in the form

$$M_{\mu}(\underline{R}, \omega, \omega_k) \sim A(\underline{R}, \omega) . \langle \psi_{\kappa} | \underline{p} | \psi_b \rangle .$$

If $\rho(\omega)$ and the matrix element $\langle\psi_{\kappa}|\underline{p}|\psi_b\rangle$ are effectively independent of over the narrow frequency band of the incident light, and if the incident wave is plane as we assumed, then (2.12b) reduces to

$$W_{\mu}(\underline{R}, t) = \{\rho(\omega_0)\}^{\frac{1}{2}} \underline{V}(\underline{R}, t) . \langle \psi_{\mu, k_0} | \underline{p} | \psi_b \rangle \tag{2.14}$$

where $\omega_0 = k_0 c$ is the mid-frequency, and $\underline{V}(R,t)$ is the vector function obtained from $\underline{A}(R,t)$ by suppressing the negative frequency components in the Fourier integral† (2.6):

---

†Such a representation of the field, obtained by suppressing the negative Fourier components is customarily employed in the classical theory of optical coherence, under the name of *analytic signal*, a concept due to D.Gabor (cf.Born and Wolf 1959). The same representation has played an important role already in the early quantum mechanical investigations of radiation and coherence based on the correspondence principle, and is also implicit in some older pioneering researches of von Laue relating to coherence and thermodynamics of light.

$$\underline{V}(\underline{R},t) = \int_0^\infty A \quad (\underline{R},\omega)e^{-i\omega t}\, d\omega. \tag{2.15}$$

From Maxwell's equation $\underline{V}(\underline{R},t)$ is transverse, i.e. normal to the direction of propagation of the wave. Let us write

$$\underline{V}(\underline{R},t) = V(\underline{R},t)\underline{\varepsilon} \tag{2.16}$$

where $\underline{\varepsilon}$ is a unit (generally complex) vector and $V(\underline{R},t)$ is a complex scalar function. Now one may readily show that for a time interval Δt, which is short compared with the reciprocal of the effective bandwidth of the light (as is here assumed),

$$\frac{1}{\Delta t}\int_{-\Delta t/2}^{\Delta t/2} V^*(\underline{R},t+\tau)V(\underline{R},t+\tau)d\tau \simeq V^*(\underline{R},t)V(R,t).$$

With the aid of this result and (2.14) and (2.16), (2.13) becomes

$$P(t) = \frac{2\pi e^2 N}{\hbar^2 m^2 c^2}\rho(\omega_0)V^*(\underline{R},t)V(\underline{R},t)\sum_\mu |\underline{\varepsilon}.\ \langle\psi_{\mu,k_0}|\underline{p}|\psi_b\rangle|^2. \tag{2.17}$$

Because $\sum_\mu$ involves summation over all possible polarizations of the electron the result will be independent of $\underline{\varepsilon}$, so that we may write (2.17) in the form

$$P(t) = \alpha\underline{V}^*(\underline{R},t).\underline{V}(\underline{R},t) \tag{2.18a}$$

where α represents the quantum efficiency of the photoelectric detector. This result does not depend in an essential way on all the simplifying assumptions made. If there is a whole range of initial electron states $|\psi_b\rangle$, the factor N in (2.17) has to be replaced by a sum over these states. On the other hand, if the electron wave functions overlap, as in a metal, the electron system has to be treated appropriately, and the calculation must be modified. Nevertheless, as long as we are dealing with plane waves falling normally on a thin photoelectric layer, a factorization of the kind embodied in equation (2.14) will still be permissible. The general form of (2.18a) will therefore remain valid, although the total cross section for the process, and therefore the constant α, will be affected.

We may identify $\underline{V}^*(\underline{R},t).\underline{V}(\underline{R},t)$ with the instantaneous intensity† $I(t)$ of the classical field, and express (2.18a) in the form

$$P(t)\Delta t = \alpha I(t)\Delta t. \tag{2.18b}$$

The probability of photoemission of an electron is therefore proportional to the classical measure of the instantaneous light intensity, defined in terms of the complex analytic signal. In the idealized case of strictly monochromatic radiation this result is, of course, well known, but its generalization to a field which exhibits arbitrary fluctuations is essential for the purposes

---

†$\underline{V}^*.\underline{V}$ is not strictly proportional to the instantaneous energy density; it may be easily shown that $\underline{V}^*.\underline{V}$ represents a short-time average of $\underline{A}^2$ taken over a time interval of a few mean periods of the light vibrations.

of the present discussion (see also Brown and Twiss 1957 a).

It should be noted that in equation (2.18) probability enters in two different ways: in the fundamental uncertainties associated with the photoelectric interaction and in the fluctuations of the radiation field itself. This, of course, is a general feature of quantum statistical mechanics (cf. Landau and Lifschitz 1958).

Equation (2.18) shows that the probability of a single photoelectric transition in a small time interval $t, t+\Delta t$ is proportional to Δt. However, we are mainly interested in the probability distribution $p(n,t,T)$ of emission of n photoelectrons in a finite time interval $t, t+T$. If the different photoelectric emissions could be considered as independent statistical events in the sense of classical probability theory, it would follow from (2.18) that (see Mandel 1959, 1963 a)

$$p(n,t,T) = \frac{1}{n!}\{\alpha U(t,T)\}^n \exp\{-\alpha U(t,T)\} \tag{2.19}$$

where

$$U(t,T) = \int_t^{t+T} I(t')dt'. \tag{2.20}$$

Actually, it is possible to see that it is legitimate to proceed in this way. Consider, for example, a system of two atoms, both of which interact with the incident radiation but do not interact with each other. The product of the unperturbed energy eigenfunctions of the two individual atoms are the energy eigenfunctions of the two-atom system. It is now possible to calculate the probability amplitude for a transition to a final state in which either one or two photoelectrons have been emitted. In particular, the probability amplitude for emission of two photoelectrons in a small time interval $(t, t+\Delta t)$ may be shown to be given by the product of two expressions of the type (2.8). When we take account of the (infinite) degeneracy of the two-atom energy eigenstates, the probability for such a transition in this time interval may then be shown to be given by a product of two factors† of the type (2.10). In any case it seems plausible to look on successive photoelectric emissions from the whole photoelectric surface considered as one system as events which are substantially independent with respect to the electron system, provided the photoelectric layer is not appreciably depopulated. Similar assumptions are implicit in the derivation of (2.19) from (2.18) by Mandel (1963 a).

Equation (2.19) refers to the photoelectric counting distribution appropriate to a single realization of the incident electromagnetic field. The average of $p(n,t,T)$ over the ensemble of the incident fields is the probability that would normally be derived from counting experiments. If, as is usually the case, $I(t)$ represents a stationary ergodic process, this average will be independent of t. If we denote the ensemble average by $\bar{p}(n,t,T)$, we have from (2.19)

$$\bar{p}(n,t,T) = \frac{1}{n!}\overline{\{\alpha U(t,T)\}^n \exp\{-\alpha U(t,T)\}} \tag{2.21}$$

†This situation must be contrasted with the situation in which a direct two-electron transition from a single atom takes place.

which, in general, will *not* be a Poisson distribution.

It should be noted that for radiation fields in some states, for example in an eigenstate of the number operator, the probability distribution of U will exhibit somewhat unusual properties, not normally encountered in classical theory. However, in view of a theorem relating to the equivalence of the semi-classical and quantum representations of light beams, established recently by Sudarshan (1963 a,b), such probability distributions can in principle nevertheless always be found.

3. STATISTICAL PROPERTIES OF THE COUNTING DISTRIBUTION

Let us next consider the variance of the counts n, recorded in time intervals of duration T. The averages of n and n^2 are given by

$$\bar{n} = \sum_{n=0}^{\infty} n\bar{p}(n,t,T) \tag{3.1}$$

$$\overline{n^2} = \sum_{n=0}^{\infty} n^2\bar{p}(n,t,T). \tag{3.2}$$

Using well-known expressions for the first two moments of the Poisson distribution (see, for example, Levy and Roth 1951), we readily find from (3.1), (3.2) and (2.19) that

$$\bar{n} = \alpha\overline{U(t,T)} \tag{3.3}$$

$$\overline{n^2} = \alpha\overline{U(t,T)} + \alpha^2\overline{\{U(t,T)\}^2} \tag{3.4}$$

so that the variance

$$\overline{(\Delta n)^2} = \overline{(n-\bar{n})^2} = \overline{n^2} - (\bar{n})^2$$

is given by

$$\overline{(\Delta n)^2} = \bar{n} + \alpha^2\overline{(\Delta U)^2} \tag{3.5}$$

where

$$\overline{(\Delta U)^2} = \overline{(U-U)^2} = \overline{U^2} - (\bar{U})^2$$

is the variance of U(t,T).

The formula (3.5) has evidently a very simple interpretation. It shows that the variance of the fluctuations in the number of ejected photoelectrons may be regarded as having two separate contributions: (i) from the fluctuations in the number of particles obeying the classical Poisson distribution (term $\bar{n}$), and (ii) from the fluctuations in a classical wave field (wave interference term $\alpha^2\overline{(\Delta U)^2}$). This result, which holds for any radiation field, is strictly analogous to a celebrated result of Einstein (1909 a,b)† relating to energy

†For a lucid account of the significance of Einstein's result, see Born (1949).

fluctuations in an enclosure containing black-body radiation, under conditions of thermal equilibrium. Fürth (1928) has later shown that the same result holds for energy fluctuations of thermal radiation of spectral compositions other than that appropriate to black-body radiation, but like Einstein's analysis, Fürth's considerations apply to closed systems only. We have now shown that a fluctuation formula of this type is also valid for counting fluctuations in *time* intervals, for any light beam (i.e. thermal or non-thermal and stationary and non-stationary), at points that may be situated far away from the sources of the light. Although the result refers to the fluctuations of the photoelectric counts, it can be regarded as reflecting the fluctuation properties of the light itself, in so far as they are accessible to measurement.

Since equation (3.5) applies to any light beam, whether of thermal origin or not, it is likely to be useful in connection with photoelectric experiments relating to fluctuations of light generated by optical masers. In this connection we note that, if an optical maser operates on a single mode and is well stabilized, then the variance $\overline{(\Delta U)^2}$ will be negligible (absence of classical wave intensity fluctuations). In this case (3.5) reduces to†

$$\overline{(\Delta n)^2} = \bar{n} \tag{3.6}$$

i.e. the variance is the same as for a system of classical particles.

We may draw some further conclusions from the results of §2. First, let us again consider the case when the light intensity of the classical wave field does not fluctuate significantly. In this case (2.21) becomes

$$\bar{p}(n,t,T) = \frac{1}{n!} \bar{n}^n \exp(-\bar{n}) \tag{3.7}$$

where

$$\bar{n} = \alpha\bar{U} = \alpha\bar{I}\,T\,. \tag{3.8}$$

Thus we see that the photoelectrons now obey the Poisson distribution. This situation may be expected to arise in the case already referred to, namely when the photoemission is triggered off by a well-stabilized, single-mode laser beam. That in this case the distribution $\bar{p}(n,t,T)$ will be Poisson's was already noted elsewhere (Mandel 1964, Glauber 1964). This result clearly shows that departure from Poisson's statistics is not a universal consequence of the Bose-Einstein statistics of light quanta as is often erroneously believed to be the case††

If, on the other hand, the light is of thermal origin, the probability distribution $\bar{p}(n,t,T)$ may be expected to be quite different. For in this case the incident field will fluctuate appreciably and its distribution will as a rule be Gaussian (van Cittert 1934, Blanc-Lapierre and Dumontet 1955, Janossy 1957, 1959). This implies (Mandel 1963 a, p.191) that for polarized thermal light

†It should be born in mind that the formula was derived on the basis of the first-order perturbation theory. When very intense maser beams are employed the effect of multiple photon interactions might have to be included.

††In this connection see the interesting discussion by Rosenfeld (1955, especially pp.77-78).

the probability distribution of the intensity is exponential:

$$p(I) = \frac{1}{\bar{I}} \exp\left(\frac{-I}{\bar{I}}\right). \tag{3.9}$$

It may then be shown from (2.21), (3.9) and (3.8) that, if T is much smaller than the coherence time of the light, the probability distribution of the photoelectrons becomes (Mandel 1958, 1959, 1963 a)

$$\bar{p}(n,t,T) = \frac{\bar{n}^{n}}{(\bar{n}+1)^{n+1}}. \tag{3.10}$$

This will be recognized as the Bose-Einstein distribution (Morse 1962). It was derived by Bothe (1927) by a somewhat similar argument long ago.

4. MULTIPLE CORRELATIONS

The one-dimensional counting distributions do not exhaust the range of application of the semi-classical theory, for the relation (2.19) can be applied to any number N of photodetectors, each situated at a different point of the radiation field. Let n_j be the number of counts registered at the jth detector in a time interval t_j, $t_j + T$. Then

$$\overline{n_1 n_2 \ldots n_N} = \sum_{n_1=0}^{\infty} \sum_{n_2=0}^{\infty} \cdots \sum_{n_N=0}^{\infty} \overline{\left\{ n_1 n_2 \ldots n_N \prod_{j=1}^{N} p_j(n_j,t,T) \right\}} \tag{4.1}$$

where the $p_j(n_j,t,T)$ $(j = 1,2\ldots N)$ are given by expressions such as (2.19). Equation (4.1) may be rewritten in the form

$$\overline{n_1 n_2 \ldots n_N} = \overline{\sum_{n_1=0}^{\infty} n_1 p_1(n_1,t_1,T) \sum_{n_2=0}^{\infty} n_2 p_2(n_2,t_2,T) \ldots \sum_{n_N=0}^{\infty} n_N p_N(n_N,t_N,T)}. \tag{4.2}$$

Now each of the sums on the right-hand side of (4.2) represents, according to a well-known property of the Poisson distribution (2.19), the parameter of that distribution

$$\alpha_j U_j(t_j,T) = \sum_{n_j=0}^{\infty} n_j p_j(n_j,t_j,T). \tag{4.3}$$

Hence, if we substitute from (4.3) into (4.2), we obtain the formula

$$\overline{n_1 n_2 \ldots n_N} = A\overline{U_1(t_1,T) U_2(t_2,T) \ldots U_N(t_N,T)} \tag{4.4}$$

where

$$A = \alpha_1\alpha_2\ldots\alpha_N \tag{4.5}$$

represents the product of the quantum efficiencies of the N detectors. Equation (4.4) shows that *the correlation of the counts registered by the N photodetectors is proportional to the correlation in the integrated intensities* (cf. (2.20))

$$U_j(t_j, T) = \int_{t_j}^{t_j+T} I_j(t')\,dt' \qquad (j = 1,2,\ldots N) \tag{4.6}$$

of the classical field at the location of the N detectors.

If we substitute from (4.6) into (4.4) and recall that $I_j(t) = \underline{V}_j^*(t).\underline{V}_j(t)$ we readily find that

$$\overline{n_1 n_2 \ldots n_N} = A \int_{t_1}^{t_1+T}\int_{t_2}^{t_2+T}\ldots\int_{t_N}^{t_N+T} \Gamma^{(N,N)}(t_1',t_2',\ldots t_N')dt_1'\,dt_2'\ldots dt_N' \tag{4.7}$$

where

$$\Gamma^{(N,N)}(t_1,t_2,\ldots t_N) = \overline{\underline{V}_1^*(t_1).\underline{V}_1(t_1)\ldots\underline{V}_N^*(t_N).\underline{V}_N(t_N)}. \tag{4.8}$$

Thus the correlation of the counts is completely expressible in terms of the 2Nth order cross-correlation function of the classical field (cf. Mandel 1964, Wolf 1963, 1964). For a stationary field this correlation function will, of course, be independent of the origin of time and if, in addition, ergodicity is assumed, it can also be expressed in the form

$$\tilde{\Gamma}^{(N,N)}(\tau_2,\tau_3,\ldots\tau_N)$$

$$= \lim_{T\to\infty}\frac{1}{2T}\int_{-T}^{T} \underline{V}_1^*(t).\underline{V}_1(t)\underline{V}_2^*(t+\tau_2).\underline{V}_2(t+\tau_2)\ldots\underline{V}_N^*(t+\tau_N).\underline{V}_N(t+\tau_N)dt \tag{4.9}$$

where

$$\tau_j = t_j - t_1 \qquad (j = 2,3,\ldots N).$$

In a quantized field-theoretical treatment the correlation $\overline{n_1 n_2 \ldots n_N}$ would be expressed in terms of the expectation value of the ordered product of the corresponding creation and annihilation operators (Glauber 1963 b, 1964). This expectation value has already been shown to be equivalent to a cross correlation of the complex classical fields (Sudarshan 1963 a,b, Mandel and Wolf 1965) and the formula (4.7) emphasizes this fact once again.

We can also convert (4.4) into a correlation formula for the fluctuations $\Delta n_j = n_j - \bar{n}_j$. By making a multinomial expansion of the product $\Delta n_1 \Delta n_2 \ldots \Delta n_N$ and applying (4.4) repeatedly, we obtain the formula

$$\overline{\Delta n_1 \Delta n_2 \ldots \Delta n_N} = A\overline{\Delta U_1 \Delta U_2 \ldots \Delta U_N} \tag{4.10}$$

where

$$\Delta U_j = U_j - \bar{U}_j .$$

The value of the correlation depends, of course, on the type of light illuminating the detectors. For light from the usual thermal sources the random process $\underline{V}(t)$ will, to a good approximation, be stationary, ergodic and Gaussian, and the correlations appearing on the right-hand side of equations (4.4), (4.7) and (4.10) can then be expressed in terms of second-order cross-correlation functions (Reed 1962). In particular, for $N = 2$, (4.10) then represents the correlation effect discovered by Brown and Twiss (1956, 1957a, b). However, from (4.10) it follows that the effect will be small when the fluctuations ΔU in the integrated classical intensity are small, as may be the case for light generated by an optical maser oscillating in a single mode.

ACKNOWLEDGEMENTS

It is a pleasure to acknowledge helpful discussion with Dr. C.L. Mehta about the subject matter of this paper.

REFERENCES

Alkemade C T J 1959 *Physica 25* 1145

Blanc-Lapierre A and Dumontet P 1955 *Rev. Opt. (Théor. Instrum.) 34* 1

Born M 1949 *Natural Philosophy of Cause and Chance* (Oxford: Clarendon Press), p.80

Born M and Wolf E 1959 *Principles of Optics* (London, New York: Pergamon Press), chap. X

Bothe W 1927 *Z. Phys. 41* 345

Brown R Hanbury and Twiss R Q 1956 *Nature 177*, 27; 1957 a *Proc. Roy. Soc. A 242* 300; 1957 b *Proc. Roy. Soc. A 243* 291

Van Cittert P H 1934 *Physica 1* 201; 1939 *Physica 6* 1129

Einstein A 1909 a *Phys. Z 10* 185; 1909 b *Phys. Z 10* 817

Fürth R 1928 *Z Phys. 50* 310

Glauber R J 1963a *Phys. Rev. Letters 10* 84; 1963b *Phys. Rev. 130* 2529; 1964 *Proc. 3rd Int. Congress on Quantum Electronics*, Eds. N. Bloembergen and P. Grivet (Paris: Dunod; New York: Columbia University Press), p.111

Janossy L 1957 *Nuovo Cim. 6* 14; 1959 *Nuovo Cim. 12* 369

Jaynes E T and Cummings F W 1963 *Proc. Inst. Elect. Electron. Engrs. 51* 89

Kahn F D 1958 *Optica Acta 5* 93

Landau L D and Lifshitz E M 1958 *Statistical Physics* (London:Pergamon Press; Reading, Mass: Addison Wesley), p.18

Levy H and Roth L 1951 *Elements of Probability* (Oxford: Clarendon Press), p.143

Mandel L 1958 *Proc. Phys. Soc. 72* 1037; 1959 *Proc. Phys. Soc. 74* 233; 1963 a *Progress in Optics 2* 181, Ed. E. Wolf (Amsterdam:North-Holland);

Mandel L 1963 b *Proc. Phys. Soc. 81* 1104; 1964 *Proc. 3rd Int. Congress on Quantum Electronics*, Eds. N. Bloembergen and P. Grivet (Paris:Dunod; New York: Columbia University Press), p.101

Mandel L and Wolf E 1963 *Phys. Rev. Letters 10* 276

Mandel L and Wolf E 1965 *Rev. Mod. Phys. 37* 231

Morse P M 1962 *Thermal Physics* (New York: Benjamin), p. 218

Purcell E M 1956 *Nature 178* 1449

Reed I S 1962 *Trans. Inst. Radio Engrs, IT-8* 194

Rosenfeld L 1955 *Niels Bohr and the Development of Physics*, Ed. W. Pauli (London: Pergamon Press)

Sudarshan E C G 1963 a *Phys. Rev. Letters 10* 277; 1963 b *Proc. Symp. on Optical Masers* (New York: Brooklyn Polytechnic Press and Wiley) p. 45

Titchmarsh E C 1948 *Introduction to the Theory of Fourier Integrals*, 2nd edn (Oxford: Clarendon Press) p. 128

Wolf E 1960 *Proc. Phys. Soc. 76* 424; 1963 *Proc. Symp. on Optical Masers* (New York: Brooklyn Polytechnic Press and Wiley) p. 29; 1964 *Proc. 3rd Int. Congress on Quantum Electronics*, Eds. N. Bloembergen and P. Grivet (Paris: Dunod; New York: Columbia University Press) p. 13

Reprinted from *Am. J. Phys.*, **44** (1976), 630–635

Some Basic Properties of Stimulated and Spontaneous Emission: A Semiclassical Approach

A. V. DURRANT
Department of Physics, The Open University, Milton Keynes MK7 6AA, UK

The coherence and directional properties of stimulated and spontaneous emission can be understood without a knowledge of quantum electrodynamics. This paper presents a semiclassical description of the interaction between a free atom and a weak optical field under conditions of either a stimulated emission or a stimulated absorption resonance. It is shown that, under certain conditions, the stimulated or spontaneous radiation fields emitted from a sample of atoms will propagate in the same direction as the incident beam, will be coherent with the incident beam, and will have the correct phase to provide amplification or attenuation.

1. INTRODUCTION

The coherence and directional properties of stimulated and spontaneous emission are usually introduced to the student in the context of the laser, and the following assertions, or something like them, are frequently made: (a) stimulated emission is coherent with the stimulating light, but spontaneous emission is not; (b) stimulated emission enters and amplifies the mode of the stimulated field. What these assertions actually mean, and the extent to which they are generally true outside the context of an optical cavity, are not usually made clear in elementary discussions. Most physicists agree that the proper theory for the treatment of such questions is quantum electrodynamics (QED). However, many undergraduate programs do not include a study of QED, and any appeal to concepts like photon creation operators and vacuum fluctuations etc. taken outside the context of a full QED treatment, is not likely to be satisfying or convincing. Now it is evident that only the wave aspects of light are referred to in the assertions, and so one is led to consider a semiclassical description: that is to say, a description in which the atoms are quantized while the optical fields are treated as classical electromagnetic waves.[1] An obvious advantage of a semiclassical treatment in the teaching context is that its basic concepts are within the grasp of students who have had first courses in electromagnetic theory and in quantum mechanics. In this paper we present a semiclassical investigation of the basic interactions that occur when a plane harmonic electromagnetic wave falls

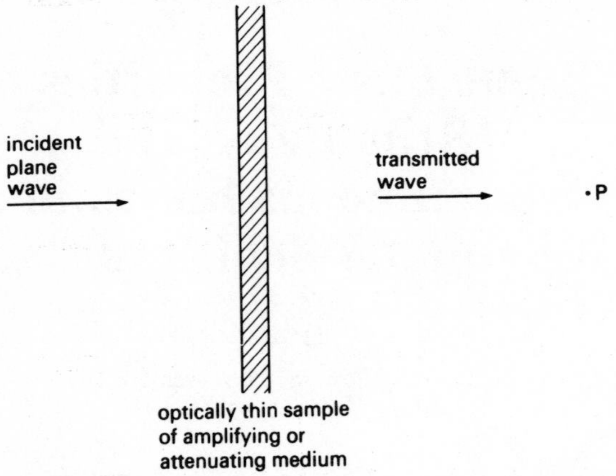

Fig.1 A plane wave of optical radiation is incident on a plane slab of atoms. The transmitted wave is a superposition of the incident plane wave and the waves emitted from the atoms in the sample. The emitted waves may consist of stimulated or spontaneous emission depending upon the detailed mechanisms of the interaction (see Fig.2).

on an optically thin sample of atoms under conditions of an absorption or an emission resonance (see Fig.1). Our conclusions will include precise statements of the coherence and directional properties of the emitted light. Our main restriction consists in taking the interaction perturbation to first order only. This corresponds to cases where the light source is weak, and leads to a linear response for the atoms. We shall make use of phenomenological decay constants to describe the spontaneous decay of excited atomic states; this involves using a non-Hermitian Hamiltonian to describe the radiating atom. This procedure is widely used and its validity well established.[2]

2. STATE OF SINGLE ATOM

We begin by calculating the quantum state of a single atom placed in an optical field whose electric vector is

$$\underline{E}_i(\underline{r},t) = \underline{e}E_0\cos(\omega t - \underline{k}\cdot\underline{r}). \tag{1}$$

Here $\underline{e}$ is a unit polarization vector, $\underline{k}$ is the propagation vector, and ω the angular frequency. $\underline{e}$ and $\underline{k}$ define the mode of the incident field. We consider first the case where the atom is in an absorption resonance with the field. The analysis may then be simply extended to the case of an emission resonance.

A. Case (i): Absorption Resonance [Fig.2(a)]

The simplest case to consider is absorption from the ground state. Let the ket $|a\rangle$ represent the ground state of the atom and let us take the zero of

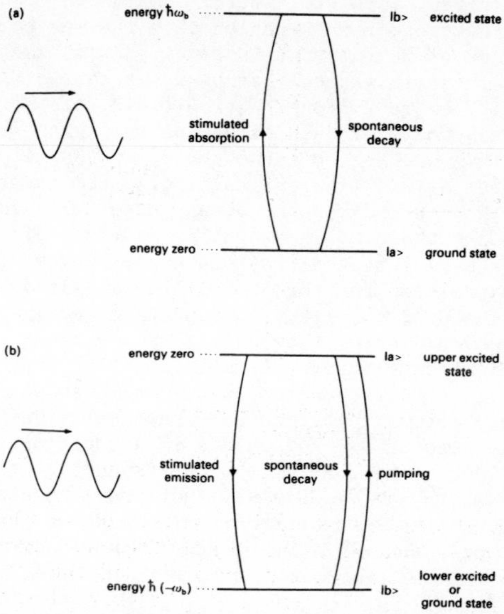

Fig.2(a) Two-level model for an absorption resonance. The incident optical field stimulates the absorption transition from the ground state |a> of an atom to an excited state |b>. In the steady state the stimulated absorption to |b> is balanced by spontaneous decay of |b> back to |a>. Note that for a weak incident field the stimulated decay of |b> is negligible compared with the spontaneous decay; the stimulated decay does not appear in the first-order treatment of the absorption process. (b) Two-level model for an emission resonance. The incident optical field stimulates the emission transition from the upper excited state |a> to a lower state |b>. In the steady state the population of state |a> is maintained against the combined effects of spontaneous and stimulated decay by a strong externally applied pumping mechanism.

energy at this level. The first excited state |b> then has energy $\hbar\omega_b$. Using the electric dipole approximation, we can write the interaction Hamiltonian as

$$\hat{H}_D = - \underline{E}_i(\underline{r}',t)\cdot\underline{\hat{D}}, \tag{2}$$

where $\underline{D}$ is the electric dipole moment operator for the atom situated at position $\underline{r}'$. We shall assume that state |b> is the only state in resonance with the field,[3] and that |a> is the only state below |b> into which spontaneous decay from |b> can occur. With these assumptions we are effectively dealing with a two-level atom, and a perturbation expansion for the state of the atom under the influence of (2) need contain only two terms.

Before we can apply perturbation theory we must first specify the unperturbed, or zero-order, states of the atom. We cannot use the states |a> and |b> as they stand because we know that, even in the absence of an applied perturbation such as (2), any excited state will suffer spontaneous decay. This decay has its origin in a much stronger interaction than the one that we are considering. This, in the language of QED, is the interaction with the vacuum fluctuations of the field; or, in semiclassical language, it is the interaction with the so-called self-field, or radiation-reaction field, of the atom.[4] For our purpose we can take account of this strong perturbation in a purely phenomenological manner. This we base on the observation that a population of excited atoms decays exponentially. Thus, if a single atom is prepared at time zero in the state $|b\rangle$, then in the absence of any externally applied fields, the probability of finding the atom in state $|b\rangle$ at time t is $\exp(-\Gamma_b t)$, where the decay constant Γ_b is just the Einstein A coefficient for the transition. The most general expression for the probability amplitude of state $|b\rangle$ at time t is therefore $\exp[-(\Gamma_b t/2 + i\delta_b)]$. Thus we take account of the spontaneous decay of state $|b\rangle$ by replacing the usual time-dependent factor $\exp(-i\omega_b t)$ by the modified time dependence $\exp\{-i[(\omega_b - i\Gamma_b/2)t + \delta_b]\}$ [the phase factor $\exp(-i\delta_b)$, which is of course unimportant when it appears as an overall multiplying factor in a quantum state, can have physical significance when it appears as a relative phase between two components in a superposition state]. In a similar way we could take account of the corresponding growth in the population of the ground state $|a\rangle$. This will not prove necessary however since we are restricting the analysis to weak optical fields so that, to first order, the ground-state population will not be significantly disturbed by the optical perturbation.

We now write a perturbation expansion for the state of the atom at time t:

$$|t\rangle = \alpha_a(t)|\alpha\rangle$$
$$+ \alpha_b(t)|b\rangle \exp\{-i[(\omega_b - i\Gamma_b/2)t + \delta_b] \quad , \tag{3}$$

with the condition $|\alpha_a(t)|^2 \approx 1$ imposed by our assumption of a weak optical field. We could of course subsume the phase factor $\exp(-i\delta_b)$ in the unknown amplitude $\alpha_b(t)$. By displaying this factor explicitly we are merely emphasizing the fact that at this stage in the argument we cannot specify the relative phase of the two components in the superposition. Our problem now is to find the induced probability amplitude $\alpha_b(t)$ of the excited state, for it is this that will determine the optical field radiated by the atom. The state $|t\rangle$ satisfies the Schrödinger equation for the Hamiltonian $(\hat{H}_0 + \hat{H}_D)$, where $\hat{H}_D$ is the interaction Hamiltonian given by (2), and $\hat{H}_0$ is the Hamiltonian of the atom in the absence of the applied optical field. $\hat{H}_0$ is assumed to contain the effects of spontaneous decay as well as the internal interactions within the atom. With our phenomenological description $\hat{H}_0$ will be non-Hermitian with eigenvalues 0 and $\hbar(\omega_b - i\Gamma_b/2)$ for the two states.

We now proceed to use time-dependent perturbation theory to find the excited-state amplitude $\alpha_b(t)$ to first order. To do this we substitute the state $|t\rangle$ of Eq.(3) into the Schrödinger equation for $(\hat{H}_0 + \hat{H}_D)$. Knowing the eigenstates and eigenvalues of $\hat{H}_0$ we obtain directly

$$\frac{d}{dt'} \alpha_b(t') = \frac{iE_0D_{ba}}{2\hbar} \alpha_a(t')$$
$$\times (\exp\{i[(\omega_b - \omega - i\Gamma_b/2)t' + \underline{k}\cdot\underline{r}']$$
$$+ \exp\{i[(\omega_b + \omega - i\Gamma_b/2)t' - \underline{k}\cdot\underline{r}'])\exp(i\delta_b)$$

at any time t'. Here $D_{ba} = \langle b|\underline{e}\cdot\hat{D}|a\rangle$. We now put $\alpha_a(t') = 1$ for all t' and integrate from $t' = -\infty$ to $t' = t$ to obtain the steady-state solution at time t:

$$\alpha_b(t) = \frac{E_o D_{ba}}{2\hbar}\left(\frac{\exp\{i[(\omega_b - \omega - i\Gamma_b/2)t + \underline{k}\cdot\underline{r}']\}}{\omega_b - \omega - i\Gamma_b/2} + \frac{\exp\{i[\omega_b + \omega - i\Gamma_b/2)t - \underline{k}\cdot\underline{r}']\}}{\omega_b + \omega - i\Gamma_b/2}\right)\exp(i\delta_b). \tag{4}$$

Since the frequencies ω_b and ω are essentially positive we can strike out the second "nonresonant" term. Then, substituting (4) into (3) we obtain the steady state of the atom:

$$|t\rangle = |a\rangle + \frac{E_o D_{ba}}{2\hbar}|b\rangle \frac{\exp[-i(\omega t - \underline{k}\cdot\underline{r}')]}{\omega_b - \omega - i\Gamma_b/2}. \tag{5}$$

We note that the angular frequency of oscillation of the excited state amplitude is equal to the optical angular frequency ω, and that the phase δ_b no longer appears. This means that the oscillations have a definite phase relation with respect to the electric vector of the incident light at the location $\underline{r}'$ of the atom. This result will have important consequences when we come to consider the radiation field emitted by the atom in Section 3.

We emphasize that Eq.(5) is a steady-state solution in which the rate of optical excitation to state $|b\rangle$ is balanced by the rate of spontaneous decay of $|b\rangle$ back to $|a\rangle$. The rate of stimulated decay from $|b\rangle$ to $|a\rangle$ is very much smaller than the rate of spontaneous decay and does not appear in this first-order treatment. An experimental situation corresponding closely to our model is the irradiation of mercury vapour with a steady beam of 2537-Å resonance radiation which excites the transition $^1S_0 \rightarrow {}^3P_1$. For an even isotope of mercury the ground state is nondegenerate as assumed in our model. The steady state is maintained by the spontaneous decay of the excited state $^3P_1 \rightarrow {}^1S_0$ with the emission of a field of resonance fluorescence, as is well known.

B. Case (ii): Emission Resonance [Fig.2(b)]

We now consider the case where the atom is in an excited state and the applied optical perturbation, Eq.(2), induces a small oscillating probability amplitude for a lower state. We shall assume, as in case (i), a two-level system driven in the steady state. In this case though, the steady-state requirement demands that we postulate a pumping mechanism, as indicated in Fig.2(b), which maintains the upper state population against the effects of spontaneous decay and the applied optical perturbation. We have not given the details of the pumping process in Fig.2(b). Optical pumping via a third higher state not shown in the figure, or excitation by collisions with energetic electrons in a discharge, or resonant exchange of excitation by collision with an excited foreign gas atom as occurs in the He-Ne laser, are possible examples of a suitable pumping mechanism. We shall make two assumptions. First, that the effect of the pumping is to produce an exponential decay of any population of atoms prepared in the lower state; and second, that the rate constant for pumping from the lower to the upper state is very much greater than the rate constant for spontaneous decay of the upper state, so the probability of finding the atom in the upper state is always nearly unity. We shall then ignore the effects of spontaneous decay on the populations of the two states. We now choose to use the symbols $|a\rangle$ and $|b\rangle$ to represent the upper and lower states, respectively, of our two-level system. With this notation we can take over Eq.(3) as the two-state perturbation expansion for the atom under the influence of the applied optical perturbation, except that ω_b in the complex exponential must now be replaced by $-\omega_b$ (with ω_b positive) if we keep the convention that energy is to be measured from state $|a\rangle$. Thus we write

$$|t\rangle = \alpha_a(t)|a\rangle + \alpha_b(t)|b\rangle \exp\{i[(\omega_b + i\Gamma_b/2)t - \delta_b]\},$$

with $|\alpha_a(t)|^2 \approx 1$ as before, but here this condition expresses the strong pumping from $|b\rangle$ to $|a\rangle$. The phenomenological decay constant Γ_b here describes the effect of the pumping mechanism in removing population from state $|b\rangle$; thus if a population of atoms were prepared in state $|b\rangle$ it would decay with rate constant Γ_b. We have put the phenomenological decay constant of state $|a\rangle$, Γ_a, equal to zero. This expresses our assumption that the population of state $|a\rangle$ is maintained against spontaneous decay by the strong pumping. We note that this phenomenological treatment hides many of the details of the pumping cycle which are not of interest to us; in particular the spontaneous decay of state $|a\rangle$ and the corresponding spontaneous emission field cannot appear in our analysis which is concerned only with the stimulated decay of $|a\rangle$ under the influence of the applied optical perturbation. We shall return to this point in Section 3.

The perturbation calculation will now proceed exactly as in case (i) except for the change in the sign of ω_b which makes the second term in Eq.(4) resonant and the first nonresonant. Equation (5) which represents the steady state of the atom for an absorption resonance may now be generalised to include the case of emission resonance:

$$|t\rangle = |a\rangle \pm \frac{E_o D_{ba}}{2\hbar}|b\rangle \frac{\exp[\mp i(\omega t - \underline{k}\cdot\underline{r}')]}{\omega_b - \omega \mp i\Gamma_b/2}. \tag{6}$$

Here the upper signs are to be taken for an absorption resonance from a ground state $|a\rangle$; and the lower signs for an emission resonance from a pumped excited state $|a\rangle$. The two cases differ only in the phase of the oscillation relative to that of the applied field, and in the interpretations of $|a\rangle$, $|b\rangle$, and Γ_b.

3. RADIATION FIELD FROM A SINGLE ATOM

The expectation value of the electric dipole moment operator $\hat{\underline{D}}$ in the state $|t\rangle$ of (6) is readily found to be

$$\langle t|\hat{\underline{D}}|t\rangle = \pm \underline{e}E_o X \cos(\omega t - \underline{k}\cdot\underline{r}' - \alpha), \tag{7}$$

with

$$X = \frac{|D_{ba}|^2/\hbar}{[(\omega_b - \omega)^2 + \Gamma_b^2/4]^{1/2}}$$

and

$$\tan\alpha = \frac{\Gamma_b/2}{\omega_b - \omega}.$$

Equation (7) describes forced simple harmonic oscillations of the induced electric dipole moment of the atom, with a frequency-dependent resonance factor X and relative phase α. We note that at the peak of the resonance, $\omega = \omega_b$, the phase of the induced moment lags the applied field by $\pi/2$ when the + sign is taken (absorption resonance), and leads by $\pi/2$ when the - sign is taken (emission resonance).

The classical expression for the electric vector of the radiation field at a point $\underline{r}$ a distance $R >> \lambda$ from an oscillating electric dipole moment $\underline{D}(t)$ situated at $\underline{r}'$ is

$$\underline{E}(\underline{r},t) = (\omega^2/4\pi\varepsilon_o c^2 R) \times [\underline{D}(t - R/c) - \underline{n}\cdot\underline{D}(t - R/c)\underline{n}], \tag{8}$$

where $\underline{n}$ is the unit vector $(\underline{r} - \underline{r}')/|\underline{r} - \underline{r}'|$, $R = |\underline{r} - \underline{r}'|$, and ω is the angular frequency of the dipole. If we now replace $\underline{D}(t)$ in Eq.(8) by the expectation value of the operator $\hat{\underline{D}}$ given by Eq.(7), we obtain a radiation field[5]

$$\underline{E}(\underline{r},t) = \pm (\omega^2/4\pi\varepsilon_o c^2 R)E_o[\underline{e} - (\underline{n}.\underline{e})\underline{n}]$$
$$\times X\cos(\omega t - \underline{k}\ \underline{r}' - \alpha - kR), \tag{9}$$

with $k = |\underline{k}| = \omega/c$. We shall refer to this radiation field as the *induced field*.

A. Coherence Properties of Induced Field

If we know the position vector $\underline{r}'$ of the atom, then the phase of the induced field is well defined. The induced field is therefore coherent with the incident field. Now this conclusion applies whether we take the + sign or the - sign in (9). In the first case we are dealing with an absorption resonance from the ground state. In our first-order theory the excited atoms return to the ground state by spontaneous decay, and the induced field represents the corresponding field of spontaneous emission. In the case of mercury vapour irradiated by 2537-Å resonance radiation we would refer to this field as the resonance fluorescence. In the second case we are dealing with an emission resonance from an excited state, and the induced field represents the field of stimulated emission associated with the stimulated decay of the excited state to the lower state. The distinction between spontaneous and stimulated emission appears here in the relative phase rather than in the fundamental coherence properties of the two processes. In what sense then are we to understand assertion (a) in Section 1 which refers to the incoherent nature of spontaneous emission? To answer this question we must examine some of the assumptions of our model.

B. Incoherent Emissions[11]

Consider first the case of absorption resonance. We have assumed in our two-state model that the ground state $|a\rangle$ is the only lower state accessible from $|b\rangle$ by spontaneous decay. If we remove this restriction and allow other such states $|d\rangle$, etc., which may or may not be degenerate with $|a\rangle$, then we must include terms like

$$\alpha_d(t)|d\rangle \exp\{-i[\omega_d - i\Gamma_d/2)t + \delta_d]\}$$

in our assumed superposition state (3). Continuing with our assumption of a weak incident field we suppose that the steady-state probabilities $|\alpha_d(t)|^2$, etc., are, like $|\alpha_a(t)|^2$, substantially unchanged by the perturbation, so that condition $|\alpha_a(t)|^2 \approx 1$ of Eq.(3) is replaced by $|\alpha_a(t)|^2 + |\alpha_d(t)|^2 + \cdots \approx 1$. The calculated expectation value of $\hat{\underline{D}}$, and therefore the emitted radiation field, will now contain additional terms representing the spontaneous emission from the upper state $|b\rangle$ to the lower states $|d\rangle$, etc. Now these additional terms will contain phase factors $\exp(-i\delta_d)$, etc. These phase factors have their origins in the detailed mechanisms of the relaxation processes responsible for maintaining the steady-state populations, and must therefore be regarded as unknown random variables as far as our analysis is concerned. We conclude that the phases of the additional spontaneous emissions are unrelated to the phase of the incident optical field, i.e., these spontaneous emissions are incoherent. This conclusion is supported by the QED treatment.[6] It is usual to refer to the incoherent spontaneous emission process as Raman scattering, while the coherent spontaneous emission in Eq.(9) is called Rayleigh scattering. We note that in our analysis the coherent Rayleigh-type scattering can occur at or near a resonance as well as in the optical frequency region well below a resonance where the familiar ω^4 intensity law applies.

Turning now to the case of an emission resonance, we obtain similar conclusions when states $|d\rangle$, etc., lower than state $|b\rangle$ are introduced, but in addition to this we must here include the much stronger source of spontaneous emission which occurs from the upper state $|a\rangle$ to all lower states of opposite parity. This emission does not appear explicitly in our analysis, which leads only to the induced field (9), since we used a phenomenological damping constant $\Gamma_a = 0$ for the upper state which expressed the effect of the pumping mechanism in maintaining any population of atoms prepared in state $|a\rangle$. In order to investigate the properties of the spontaneous emission from $|a\rangle$ we must go beyond the phenomenological treatment and consider the details of the pumping cycle in which individual atoms undergo spontaneous emission from state $|a\rangle$ to state $|b\rangle$ and are then pumped back to $|a\rangle$. Now with our assumption of a weak incident field the pumping cycle will be uninfluenced by, and independent of, the incident field. There will therefore be no definite phase relationship between the spontaneous emission field from $|a\rangle$ and the incident field: the fields are incoherent with one another. We note that for the weak incident fields assumed in our analysis the spontaneous emission from state $|a\rangle$, for a single atom, will be very much stronger than the induced emission from state $|a\rangle$; although, as we shall see below, the coherence property of the latter can lead to a "beaming" of the induced emissions from an assembly of atoms so that at large distances the induced field is the dominant one.

C. Directional Properties of the Induced Field

The induced field propagates outwards from the atom with an angular distribution given by the dipole factor in square brackets in Eq.(9). This is to be compared with the incident field which propagates in a single plane-wave mode of wave vector $\underline{k}$. When fields are confined to an indefinitely large volume, or indeed to a volume V of macroscopic dimensions, the mode density for plane waves, $\rho(\omega)$, takes the form of a quasicontinuum: $\rho(\omega) = V\omega^2/(\pi^2c^3)$, as is well known. This means that if we expand the dipole-like induced field in plane wave modes we will find only an infinitesimal fraction of the power in the particular mode of the incident field. How then can an induced field of stimulated emission lead to amplification of the incident light? The question becomes more acute when we examine the phase of the induced field at points $\underline{r}$ in the forward direction from the atom, i.e. for $(\underline{r}-\underline{r}')$ in the direction of $\underline{k}$. We can then put $\underline{n}\cdot\underline{e}=0$ and $\underline{k}\cdot\underline{r}'+kR=\underline{k}\cdot\underline{r}$ in Eq.(9) to obtain the forward field:

$$\underline{E}_f(\underline{r},t)=\pm\,(\omega^2/4\pi\varepsilon_o c^2 R)E_o\underline{e}X\,\cos(\omega t-\underline{k}\cdot\underline{r}-\alpha). \tag{10}$$

At the peak of an emission resonance we take the minus sign and put $\omega=\omega_b$ to give $\alpha=\pi/2$. The forward emission is therefore not in phase with the incident light; it leads by $\pi/2$!

To see how amplification of the incident light can occur we shall have to investigate the interference effects between the induced emissions from different atoms in a sample.

4. INDUCED FIELD FROM A SLAB

Here we investigate the resultant induced field emitted by the atoms in a thin slab of material placed at right angles to the direction of propagation of an incident plane wave. This is the geometry normally considered in refractive index theory. We shall first consider the idealised case of a plane wave falling on an infinite slab of atoms, and then go on to consider the physically real case of an aperture-limited plane wave falling on a finite slab.

A. Infinite Slab

Consider an optically thin infinite plane slab of noninteracting atoms placed normal to the incident light (Fig.1). We shall assume that the atoms are stationary and uniformly distributed in the slab. Let N be the number of atoms per unit volume. The induced field at a point P is found by summing the contributions from all atoms in the slab. This leads to a very well-known interference problem which was first solved by Fresnel for scalar waves using this method of half-period zones. We do not give the details here for the method is described in many standard textbooks.[7] The result can be stated as follows: the resultant field at a point P a distance $R>>\lambda$ from the slab of thickness d is[8]

$$\underline{E}_p(\underline{r},t)=[i]RN\lambda d\;\underline{E}_f(\underline{r},t).$$

$\underline{E}_f(\underline{r},t)$ is the field at P produced by forward emission from a single atom in the slab. This is given by Eq.(10). The symbol [i] denotes a phase lag of $\pi/2$. Thus we can write

$$\underline{E}_p(\underline{r},t)=\pm\,\underline{e}E_o(N\pi d/\varepsilon_o\lambda)X\times\cos(\omega t-\underline{k}\cdot\underline{r}-\alpha-\pi/2). \tag{11}$$

Equation (11) has the form of a plane wave propagating in the same plane wave mode as the incident wave. At the peak of a resonance we have $\alpha=\pi/2$ and the stimulated emission (take the - sign) is now in phase with the incident light. We can now understand how light amplification occurs: the dipole-like induced fields from all atoms in the slab interfere to construct a forward-propagating plane wave of just the right phase to provide amplification of the incident stimulating beam.[9] Similarly for an absorption resonance (take the + sign) we obtain a forward-propagating plane wave of spontaneous emission which, at the peak of the resonance, is in antiphase with the incident light, and this will give rise to attenuation. These considerations lead in an obvious way to a refractive index for the medium.

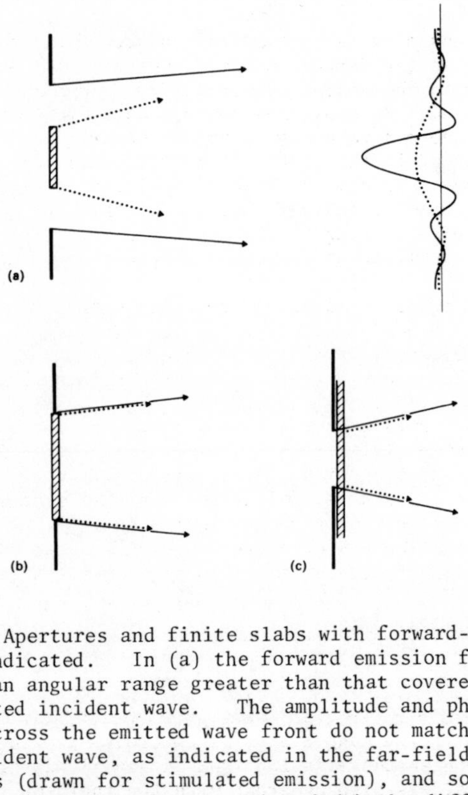

Fig. 3 Apertures and finite slabs with forward-diffraction cones indicated. In (a) the forward emission from the slab enters an angular range greater than that covered by the diffracted incident wave. The amplitude and phase variations across the emitted wave front do not match those of the incident wave, as indicated in the far-field amplitude patterns (drawn for stimulated emission), and so the amplification is distorted. In (b) and (c) the diffraction cones (nearly) coincide, and undistorted amplification or attenuation can occur.

B. Finite Slab

In physically realizable situations we have to deal with finite slabs and aperture-limited plane waves as in Fig.3(a). The interference problem for the emitting atoms then becomes, essentially, a Fresnel diffraction problem determined by the boundaries of the slab, and the idea of a forward direction for emission has to be replaced by the idea of a diffraction cone of forward directions for the slab. Similarly, the incident plane wave is spread over a range of plane wave modes due to diffraction at the apertures, and the idea of a forward direction for the incident wave has to be replaced by the idea of a diffraction cone covering a finite range of plane wave modes. Undistorted amplification (or attenuation) of the diffraction-limited incident wave can now occur only if the two diffraction problems coincide; that is to say, if the forward diffraction cone for the slab coincides with the forward diffraction cone defined by the apertures. See Figs.3(b) and 3(c). Such coincidence is found in a laser when the amplifying medium fills the whole of the optical cavity.

C. Incoherence in the Induced Field

Within the assumptions of this section the induced field is completely coherent and propagates in the forward modes. One of these assumptions, which was made implicitly in the acceptance of Eq.(10), concerns the description of the emitting medium: we took the atomic density to be uniform and continuous. In view of the discrete nature of any material medium, and the random atomic motions inherent in the state of thermal equilibrium, we should regard the atomic density as a random function of position and time, and the symbol N that we have used, as the ensemble average of this function. The fluctuations about the ensemble average give rise to an incoherent component of the induced field,[10] in addition to the coherent one that we have calculated from the ensemble average. The incoherent component is not "beamed" into the forward modes but appears in lateral directions with a power which is proportional to the total number of atoms present in the sample, and which, for an absorption resonance, provides the power balance to compensate for the attenuation of the incident beam. From a microscopic point of view the origin of the incoherence can be found from inspection of Eq.(9) which reveals that for emission directions $\underline{n}$ outside the forward direction, the phase is not independent of the random position $\underline{r}'$ of the emitting atom. From a macroscopic point of view we are dealing with the scattering of the incident light by density fluctuations as we have indicated. This is a well-developed subject which we do not need to pursue here, but which is relevant to our conclusions; for we have now identified a source of incoherence in addition to the Raman scattering and related processes at the quantum level which we discussed in Section 3.

5. SUMMARY

Stimulated and spontaneous emissions have the following properties.[11] The stimulated emission from a single atom is coherent with the incident stimulating beam. The spontaneous emission from an atom in absorption resonance with an incident field is also coherent with the incident field provided the atom returns to its original quantum state. Other spontaneous emissions are incoherent. For electric dipole transitions the coherent emission from a single atom, whether stimulated or spontaneous, has a dipole-like angular distribution. For coherent emission from a macroscopic sample, the dipole-like angular distribution can be drastically modified by interference effects. In particular, when the diffraction limiting of the incident light coincides with the boundaries of the sample, the coherent emissions enter the mode or modes of the incident light allowing amplification by stimulated emission or attenuation by absorption and spontaneous emission. Thermal density fluctuations in the sample degrade the coherence of the emitted light and give rise to a lateral dipole-like field which is incoherent with the incident beam.

REFERENCES

1. Semiclassical theory has of course been widely used in the theory of the

laser as in, for example, the classic paper by W.E. Lamb, Jr., 1964 *Phys. Rev. 134* A1429. In most of these works the medium is described by a macroscopic polarization obeying Maxwell's equations, and the detailed mechanisms by which coherence and directional properties are established and hidden in the formalism.

2. The use of damping Hamiltonians is discussed by R. Loudon, *The Quantum Theory of Light* (Clarendon, Oxford, 1973), Chap. 8; and by W. Louisell, *Radiation and Noise in Quantum Electronics* (McGraw-Hill, New York, 1964) Chap. 7.

3. Magnetic degeneracy of the upper level does not cause difficulty for we can regard the state $|b\rangle$ as that linear superposition of the degenerate magnetic substates which can be excited from the ground state by the polarized electric vector of the incident light.

4. The self-field of an accelerating electron is discussed in a number of textbooks on classical electrodynamics, for example, W.K.H. Panofsky and M. Phillips, *Classical Electricity and Magnetism* (Addison-Wesley, Reading, PA, 1964), Chap. 21. The mechanism by which the self-field can cause spontaneous decay of excited atomic states had been investigated by G.W. Series, in *Optical Pumping and Atomic Line Shape*, edited by T. Skalinsky (Panstwowe Wydawnictwo Naukowe, Warszawa, 1969), pp. 25-41. See also papers by W.L. Lama and L. Mandel and by R.K. Bullough, in *Coherence and Quantum Optics*, edited by L. Mandel and E. Wolf (Plenum, New York, 1973).

5. Semiclassical electrodynamics has traditionally used a number of techniques, based on the correspondence principle, for calculating the radiation field emitted by an atom. See, for example, E.U. Condon and G.H. Shortley, *The Theory of Atomic Spectra* (Cambridge University, Cambridge, 1963), Secs. 4 and 5. Our use of the expectation value of $\hat{\underline{D}}$ has a strong heuristic appeal and, within the assumptions of this section, leads to the same results as the other techniques. This topic is the subject of current research in the field of semiclassical electrodynamics. See papers under the heading: "Quantum Electrodynamics and Alternative Theories, I, II, and III," in *Coherence and Quantum Optics*, edited by L. Mandel and E. Wolf (Plenum, New York, 1973).

6. See W. Heitler, *The Quantum Theory of Radiation* (Clarendon, Oxford, 1966), Chap. V.

7. See, for example, R.W. Ditchburn, *Light* (Blackie, London, 1963), Chap. 6. See also R. de L. Kronig, 1926 *J. Opt. Soc. Am. 12* 547.

8. This result can be justified for vector wave fields by direct integration of the induced fields from all atoms in the slab.

9. The interference problem that we have described appears also in the QED treatment where the corresponding matrix element is summed over the contributions of the different atoms before it is squared to obtain the total transition rate. The procedure is illustrated in R. Loudon *The Quantum Theory of Light* (Clarendon, Oxford, 1973), pp. 319-321. Although his treatment refers specifically to third-harmonic generation, the procedure for summing over all atomic positions is quite general for coherent emission processes, provided the appropriate expression for $\Delta\underline{k}$ (his notation) is used.

10. This is the scattering of light by density fluctuations; see, for example, N.G. van Kampen, *Quantum Optics*, edited by R.J. Glauber (Academic, New York, 1969).

11. At the risk of repetition, we emphasize that our treatment of absorption

from a ground state $|a\rangle$ is restricted to weakly excited systems ($|\alpha_a(t)|^2 \approx 1$). When the optical excitation is strong, the coherent field that we have calculated using $\langle t|\underline{D}|t\rangle$ does not give the complete spontaneous emission field. I.R. Senitzky [1958 *Phys. Rev. 111* 3] has shown that the spontaneous emission from a strongly excited two-level system contains an incoherent component as well. In the limit of weak excitation, however, the intensity of this incoherent component is negligible compared with that of the coherent field.

Reprinted from *J. Phys.*, **B7** (1974), L198–202

Observation of the Resonant Stark Effect at Optical Frequencies*

F. SCHUDA**, C. R. STROUD, Jr.
and M. HERCHER
Institute of Optics, University of Rochester, Rochester, NY 14627, USA

The spectrum of resonantly scattered light at right angles to a sodium atomic beam is reported. The light source was a cw dye laser tuned to resonance with a hyperfine component of the D_2 line, and incident at right angles to the atomic beam. The spectrum, with the Stark effect sidebands, was recorded as a function of both the laser intensity and its detuning from resonance. The overall resolution is better than 20 MHz.

In recent years there have been a large number of attempts to predict the spectrum of the light scattered by an isolated atom illuminated by a monochromatic field tuned accurately to resonance with a two-level transition. These theoretical calculations have variously predicted a sharp line scattered spectrum (Heitler 1954), a Lorentzian with a hole burned in the middle (Bergmann 1967, Chang and Stehle 1971), a three peaked spectrum (Apanasevich 1964, Newstein 1968, Mollow 1969, Stroud 1971, 1973), and even more complex shapes (Morozov 1969, Gush and Gush 1972). There have been some experimental studies in this area (Hertz et al 1968, Hänsch et al 1969, Shahin and Hänsch 1973), but none have yet had the resolution to provide an unambiguous answer to this problem. In this letter we will describe an experiment which measures this spectrum in detail and determines that over a wide range of incident field intensities, the spectrum has three peaks.

This structure in the scattered spectrum has been termed the dynamic, or AC, Stark effect. The Stark effect produced by an optical field is dramatically enhanced by tuning the field to within the natural linewidth of a two level transition. We have taken advantage of the precise tunability of a dye laser, as well as the well defined resonance frequency of a collimated atomic

*This research was supported in part by the NSF and by the Advanced Research Projects Agency under Contract No. DAHCO 4-71-C-0046.

**Present address: Spectra Physics, 1256 West Middlefield Road, Mountain View, California 94040, U.S.A.

beam illuminated at right angles, to maximize the observed effect and make detailed measurements possible.

The usual heuristic explanation of the dynamic Stark effect is that it is just the analogue in frequency space of optical nutation. We have found, however, that we are unable to explain our experiments in terms of the conventional phenomenologically damped semiclassical equations usually used to describe nutation (Tang and Silverman 1966, Hocker and Tang 1968, Hoff et al 1970, Brewer and Shoemaker 1973). Thus the present experiment serves to point up limitations of phenomenological semiclassical theory as well as determining the nature of the spectrum.

The apparatus for observation of the resonant effect at 5890 Å in sodium is similar to that described in an earlier publication (Schuda et al 1973). A dye laser beam and an atomic beam are crossed at right angles and the resonant fluorescence from the interaction region is analysed. Figure 1 shows the experimental set-up including intensity monitoring apparatus and Fabry-Perot interferometer.

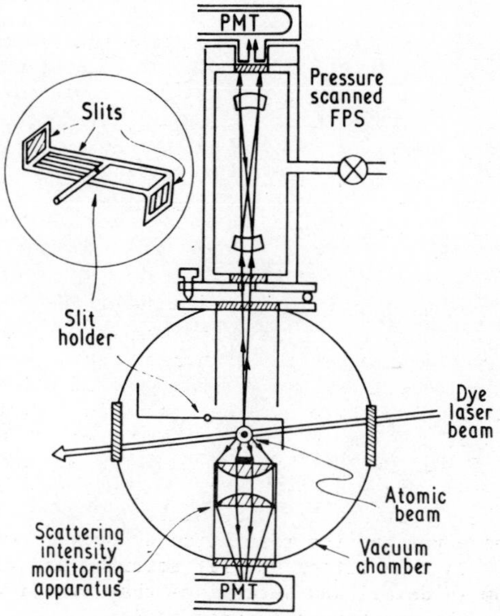

Fig. 1. Experimental apparatus for AC Stark effect observation. The atomic beam is travelling out of the plane of the paper.

The cw dye laser is first coarse-tuned to resonance in a sodium absorption cell. The laser is then directed to intersect the atomic beam and fine-tuned to the hyperfine transition F = 2 → 3 of the sodium D_2 line by monitoring the resonance scattering from the interaction region. This particular atomic transition was chosen because it is a two-level transition and the atomic population cannot be optically pumped into a third level (see Fig. 2a). The dye laser is stable to within ± 1.5 MHz for a period of 10 seconds so that during a single 5 s spectral scan the laser frequency remains essentially stationary relative to the 10 MHz natural linewidth.

Upon entering the chamber the laser beam is apertured by a slit shown in Fig. 1. The data presented here was taken with the 1 mm wide slit so that the atoms coming down the atomic beam are illuminated over a 1 mm region; the corresponding transit time is approximately 100 natural lifetimes. The laser beam electric field vector is parallel to the atomic beam. The atomic beam is collimated to 4 milliradians by passing through two 0.5 mm circular apertures which are 25 cm apart. (The reason for the relatively large apertures is to obtain a large atomic flux and therefore a large scattered intensity.) The dye laser and atomic beam are perpendicular to within 2 milliradians, so that the residual Doppler width in excitation is no more than 5 MHz.

Scattered light from the interaction region can go into one of two collecting systems. The scattered intensity monitoring system merely collects the scattered light within a large solid angle and indicates the location of the laser frequency relative to the centre of the absorption line. The distance from the centre frequency is determined from our knowledge of the relative scattering cross section as a function of input power and detuning.

Light scattered from the atomic beam also enters the confocal Fabry-Perot interferometer on the top of the vacuum chamber. A blackened piece of foil covers the central area of the intensity-monitoring lenses so that stray light is not reflected into the Fabry-Perot. The interferometer is followed by an aperture which allows only light scattered at right angles to the atomic beam to be detected, while at the same time restricting the instrumental width of the Fabry-Perot to less than 20 MHz, thus allowing the sidebands on the emission spectrum to be well resolved. The Fabry-Perot, described by Hercher (1968), has a free spectral range of 750 MHz and is scanned in 5 seconds. The output of the photomultiplier following the Fabry-Perot drives the vertical input to an oscilloscope while the horizontal time base is set at 1 s cm^{-1}. Photographs of the traces are taken and the data is later reduced to the form shown in Figs. 2 and 3. A typical oscilloscope trace is also included in Fig. 3.

On the receding axes of Figs. 2 and 3 are plotted the laser detuning from the absorption line centre. The emission spectra for various detunings are plotted horizontally. The horizontal scale is centred on the laser frequency and not on the absorption line centre. We have observed that the central peak in every case coincides with the laser frequency to within less than 20 MHz. This conclusion was reached by simultaneously observing the spectra of both the scattered light and a portion of the incident laser beam.

The receding scale in Figs. 2(b) and 3(b) is calibrated by monitoring the scattered intensity. If the spectrally-integrated scattered intensity is plotted as a function of the frequency of the applied optical field one gets the power-broadened absorption spectrum. However, in this case, for the relatively high input intensities of 406 mW cm^{-2} and 174 mW cm^{-2}, the curve is asymmetric. This is due to another hyperfine transition at a frequency just 60 MHz lower than the $F = 2 \rightarrow 3$ transition (see Fig. 2(a)). This other transition is a three-level transition and at the 406 mW cm^{-2} power density these two transitions are mixed. The three-level transition provides a path for the escape of the atomic population from the two-level system by optical pumping. Thus, on the low frequency side of the $F = 2 \rightarrow 3$ transition, the scattering falls off much more rapidly with frequency than on the high frequency side. This effect is indicated in Fig. 2(b) by the skewed envelope which fits the peaks of the spectra going along the receding axis.

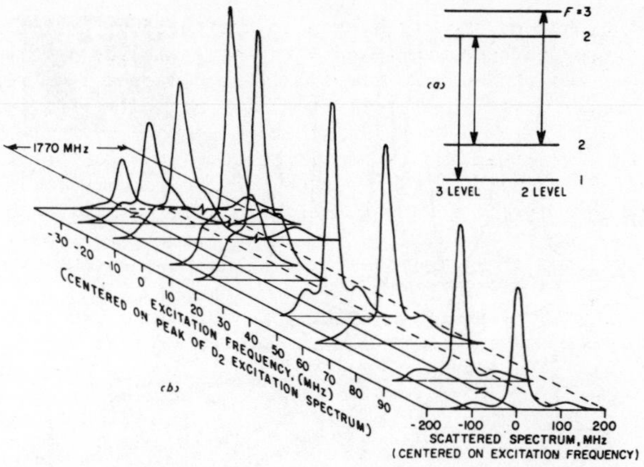

Fig. 2(a) The D_2 transitions excited by the dye laser. F = 3 and F = 2 are hyperfine components of the $3^2P_{3/2}$ level and are separated by 60 MHz. F = 2 and F = 1 of the ground $3^2S_{1/2}$ level are separated by 1772 MHz. The 2 and 3 level systems arise from the $\Delta F = 0, \pm 1$ selection rules. (b) Composite graph of the spectra of the scattered light for various laser detunings. The laser intensity is 406 mW cm^{-2} ± 100 mW cm^{-2}. The measurements of excitation frequency are accurate to within ± 7 MHz, and the scale on the scattered spectra is accurate to ± 10 MHz.

The presence of this three-level system is further verified by noting in Figs. 2 and 3 the frequency component at + 1770 MHz on the high frequency side of the spectrum when the laser is tuned below the resonance frequency. The laser partially excites the three-level transition and two frequencies are radiated, one at the laser frequency and another 1770 MHz away when an atom is optically pumped into the third level.

The spectra in the centre and lower right of Figs. 2 and 3 show quite clearly the sidebands generated by Stark splitting. As the laser intensity is increased, the sidebands are spread further from the central peak. In addition, one can see that as the detuning is increased, the sidebands spread out.

If we attempt to explain these results using phenomenological semiclassical equations (optical Bloch equations), we find that they predict that the atoms will suffer optical nutations for a few radiative lifetimes after they enter the interaction region and then they will settle down to the saturation limit with their dipole moments oscillating harmonically at the applied field frequency (Tang and Silverman 1966, Hocker and Tang 1968, Hoff et al 1970, Brewer and Shoemaker 1973). In our experiments each atom interacted with the field for approximately 100 radiative lifetimes. By using the slit of Fig. 1 we have determined that the spectrum of the light scattered by the front half of the illuminated region is identical to that from the second half. We do not observe a sharp line spectrum even for light scattered from atoms which have been in the laser beam for 50 lifetimes or more. Thus it would appear that the optical Bloch equations are inadequate for this

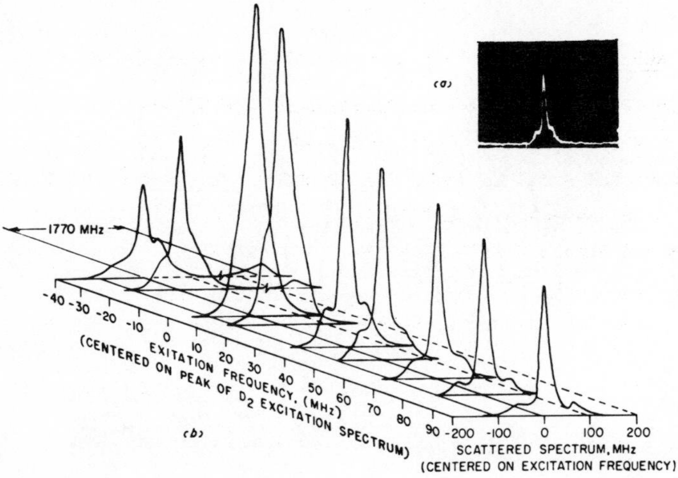

Fig. 3(a) Typical oscilloscope trace of emission spectrum. (b) Composite graph of the spectra of the scattered light for various laser detunings. The laser intensity is 174 mW cm^{-2} ± 50 mW cm^{-2}.

problem. Because of the importance of spontaneous emission in this problem, some of the theoretical attempts to predict the spectrum have used quantum electrodynamics (Stroud 1973). Unfortunately, because of the difficulty of the QED calculations, they have been carried out only for the initial transient regime; the long term solutions are unknown. There may or may not be a 'saturation steady-state limit' as in semiclassical theory (Brewer and Shoemaker 1973). The results of these calculations predict a spectrum qualitatively similar to that measured above but differing in some respects. In particular the predicted ratio of the heights of the central peak to the sidebands is 2:1 at resonance (Stroud 1973). Secondly, the areas under the sidebands are predicted to differ from each other appreciably for the detunings studied, whereas experimentally they appeared nearly equal when the laser was tuned to the high frequency side of the transition. These discrepancies are possibly due to our failure to include the nearby three-level transition in the theoretical calculations. We are carrying out a more detailed analysis and will describe our results shortly in an extended paper. One other theoretical complication has been investigated and found not to affect the predicted results. The two resonant levels are made up of 12 degenerate Zeeman sublevels. (We have experimentally cancelled the earth's field so that these levels are very accurately degenerate.) There are a number of slightly differing dipole matrix elements between various pairs of Zeeman sublevels, but they differ by too little to lead to resolvable details in our experiment.

Though the theoretical analysis of this experiment is not yet complete, we are able to reach two conclusions: (i) The spectrum of a resonantly scattered monochromatic field consists of a central peak at the applied field frequency and two symmetrically placed sidebands, and (ii) This spectrum persists when phenomenological semiclassical theory predicts that the dipole moment should have settled down to a harmonic oscillation.

REFERENCES

Apanasevich P A 1964 *Optics and Spectroscopy 16* 387-8

Bergmann S M 1967 *J. Math. Phys. 8* 159-69

Bonch-Bruevich A M and Khodovoi V A 1968 *Sov. Phys. Usp. 10* 637-57

Brewer R G and Shoemaker R L 1973 *Phys. Rev. Lett. 27* 631-4

Chang C S and Stehle P 1971 *Phys. Rev. A 4* 641-61

Gush R and Gush H P 1972 *Phys. Rev. A 6* 129-40

Hänsch T, Keil R, Schabert A, Schmelzer Ch, and Toschek P 1969 *Z Phys. 226* 293-6

Heitler W 1964 *Quantum Theory of Radiation* 3rd ed. (London, Oxford University Press).

Hercher M 1968 *Appl. Opt. 7* 951-66

Hertz J H, Hoffman K, Brunner W, Paul H, Richter G and Steudel H 1968 *Phys. Lett. 26A* 156-7

Hocker G B and Tang C L 1968 *Phys. Rev. Lett. 21* 591-4

Hoff P W, Haus H A and Bridges T J 1970 *Phys. Rev. Lett. 25* 82-4

Mollow B R 1969 *Phys. Rev. 188* 1969-75

Morozov V A 1969 *Optics and Spectroscopy 26* 62-3

Newstein M C 1968 *Phys. Rev. 167* 89-96

Schuda F, Hercher M and Stroud C R Jr. 1973 *Appl. Phys. Lett. 22* 360-2

Shahin I S and Hänsch T 1973 *Opt. Comm. 8* 312-15

Stroud C R Jr. 1971 *Phys. Rev. A 3* 1044-52

——— 1973 in *Coherence and Quantum Optics, Proc. 3rd Rochester Conf.* ed. L Mandel and E Wolf (New York: Plenum Press) 537-46

Tang C L and Silverman B D 1966 *Physics of Quantum Electronics*, ed. P Kelley, B Lax and P Tannenwald (New York: McGraw-Hill)

Reprinted from *Phys. Bull.*, **27** (1976), 21–23

Photon Bunching and Antibunching

R. LOUDON
Department of Physics, University of Essex, Wivenhoe Park, Colchester, UK

If a beam of light from an ordinary light source is allowed to fall on a photomultiplier tube which registers the arrival of photons, the remarkable phenomenon of photon bunching can be observed. That is, the photons are detected by the phototube not at randomly distributed times but in the form of clusters or bunches. The photon bunching effect first received widespread attention in the 1950s as a result of the experiments of Hanbury-Brown and Twiss (reviewed by Hanbury-Brown 1974).

The invention of the laser in the early 1960s led to a debate concerning the existence of bunches of photons in laser light. In fact, the coherent light from a laser operating well above threshold does not show the bunching effect, and the times of detection of photons by a phototube are randomly distributed. The photon bunching is a property of the incoherent or chaotic light from conventional sources.

Interest in the bunching phenomenon has again come to life with the appearance of suggestions that *antibunched* light can be generated by nonlinear interactions of laser light with matter. The observation of antibunched light is of interest for theoretical physics because its existence is only compatible with the quantum theory of light and is at variance with the classical theory.

Classical Theory

Consider first the classical theory of light. Figure 1a shows the variation with time at a fixed position of the cycle-averaged intensity $\bar{I}(t)$ of a beam of polarized light from a conventional source. Such light is termed chaotic; its intensity shows large statistical fluctuations as a result of the independence of the different source atoms which sometimes radiate constructively to give a peak, and sometimes destructively to give a trough. The time scale of the fluctuations is determined by the coherence time τ_c of the source; for a collision-broadened emission line, τ_c is the mean time between the collisions of a particular atom. The frequency spread of the light is of order $1/\tau_c$.

The properties of the cycle-averaged intensity for chaotic light can be treated statistically by a form of random-walk theory (see for example Loudon 1973). It can be shown that the mean of the *m*th power of the cycle-averaged intensity is

$$\langle \bar{I}(t)^m \rangle = m!\ \bar{I}^m \qquad (1)$$

where the angle brackets denote averages throughout, and $\bar{I}$ is the mean of the intensity over a period much longer than τ_c. The m! factor in this result is a characteristic feature of the intensity fluctuations of chaotic light.

Now consider a light beam with no intensity fluctuations, represented by a classical electric field varying sinusoidally with time and with fixed amplitude and phase. The light from single mode lasers operated well above threshold is a close experimental realization of this classical ideal, which is termed coherent light. The cycle averaged intensity of coherent light has a constant value $\bar{I}$, and trivially,

$$\langle \bar{I}(t)^m \rangle = \bar{I}^m\ . \qquad (2)$$

The averaging is of course unnecessary in this case since the quantity to be averaged is constant. The character of a laser beam changes continuously from chaotic to coherent as the pumping rate is increased from below to well above theshold.

The Hanbury-Brown and Twiss experiment shown schematically in figure 2 was the first to detect the time dependent intensity fluctuations in chaotic light. The principles of the experiment are very simple. The beam from a light source is split into two parts by a semitransparent mirror. The intensities of the separate parts are measured by two phototubes and the results are correlated and averaged electronically.

According to classical theory, the half silvered mirror splits the original light into two identical beams which show exactly the same fluctuations in cycle averaged intensity. Let $\bar{I}(t)$ be the cycle averaged intensity of each of the two split beams and let each have a long-period mean intensity $\bar{I}$. The experiment was designed to produce the mean square deviation of $\bar{I}(t)$ by taking products of the readings of the two phototube detectors. The quantity determined is thus proportional to

$$\langle \bar{I}(t)^2 \rangle - \bar{I}^2 = \begin{cases} \bar{I}^2 & \text{for chaotic light} \\ 0 & \text{for coherent light} \end{cases} \qquad (3)$$

where equations (1) and (2) have been used. The positive Hanbury-Brown and Twiss correlation of chaotic light (conventional source) is a manifestation of its intensity fluctuations of the kind shown in figure 1a. The constant-intensity coherent beam (single mode laser well above threshold) generates zero correlation. The finite resolving times in any practical experiment reduce the measured correlation for chaotic light, but the effect remains detectable (Hanbury-Brown 1974).

Another characterization of the intensity fluctuation properties of different light beams is the degree of second order coherence defined as

$$g^{(2)} = \langle \bar{I}(t)^2 \rangle / \bar{I}^2\ . \qquad (4)$$

This definition gives

$$g^{(2)} = \begin{cases} 2 \text{ for chaotic light} \\ 1 \text{ for coherent light} \end{cases} \tag{5}$$

where equations (1) and (2) have been used.

Quantum Theory

The quantum theory regards a beam of light as a stream of photons, and a measurement of the beam intensity as a counting of the arrivals of photons at a phototube. The use of photon counting experiments to examine the statistical fluctuation properties of light beams has become well established in recent years (Cummins and Pike 1974).

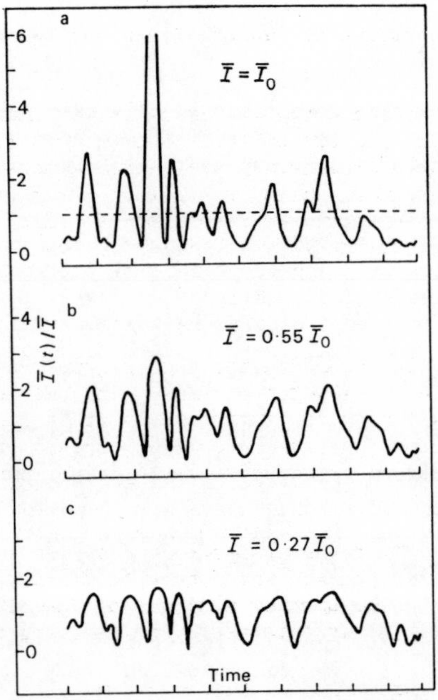

Fig. 1. *a* Intensity fluctuations of chaotic light, *b* and *c* the same section of beam after increasing amounts of two photon absorption (*after Weber 1971*). The horizontal scale divisions indicate the coherence time τ_c.

The classical intensity fluctuations of chaotic light shown in figure 1a translate into the quantum picture as the photon bunching effect. The arrival of a high intensity fluctuation at the detector generates closely spaced photon counts, while the arrival of an intensity trough produces very few photon counts. The extent in time of a photon bunch is of the order of the coherence time τ_c.

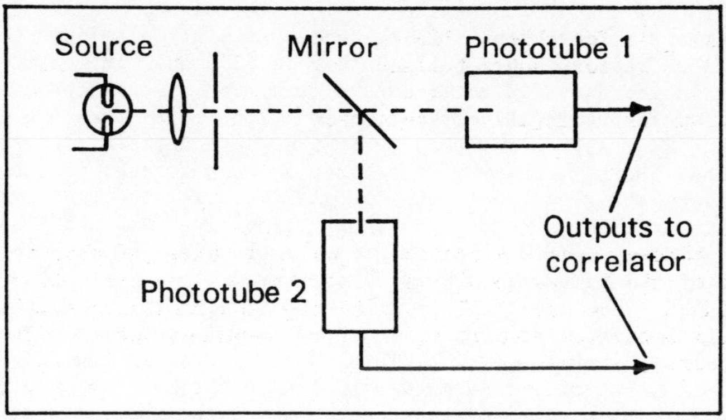

Fig. 2. Experiment of Hanbury-Brown and Twiss.

In photon language, the two detectors in a Hanbury-Brown and Twiss experiment count the number of photons transmitted through or reflected from the mirror. Consider a series of measurements in which the numbers n_1 and n_2 of photons counted by the two detectors during a constant brief time interval are recorded. Each detector registers the same average number $\bar{n}$ of counts if the series of identical experiments is sufficiently long,

$$\langle n_1 \rangle = \langle n_2 \rangle = \bar{n} \tag{6}$$

The numbers of counts in each detector in a single run are *not* however in general the same. In the quantum picture, each incident photon *either* passes through the half silvered mirror *or* is reflected from it. The two split beams are therefore not exact replicas of each other; it is only their average properties over an extended series of experiments which are identical (see also table 1 below).

The average value $\langle n_1 n_2 \rangle$ of the product of counts in each single run, the correlation function, determines the extent to which coincident counts occur in the two detectors. The quantum analogue of the classical degree of second-order coherence (4) is obtained by normalizing the correlation function,

$$g^{(2)} = \langle n_1 n_2 \rangle / \bar{n}^2 \quad . \tag{7}$$

and the quantity (3) determined in the Hanbury-Brown and Twiss experiment can be recast in the form

$$\langle n_1 n_2 \rangle - \bar{n}^2 = (g^{(2)} - 1)\bar{n}^2 \; . \tag{8}$$

The values of the quantum theory degree of second order coherence (7) for chaotic and coherent light can be obtained quite easily (see for example Loudon 1973). They are exactly the same as the results (5) given by classical theory. Despite the qualitative differences between the two pictures, classical and quantum theory produce the same predictions for experiments on the fluctuation properties of chaotic and coherent light.

The value of 2 for the degree of second order coherence of chaotic light is a manifestation of the photon bunches. In terms of the Hanbury Brown and Twiss experiment, it corresponds to the occurrence of coincident counts at twice the random rate appropriate to uncorrelated counts in the two detectors. The correlation decreases if the readings of n_1 and n_2 are separated by a time delay τ. The correlation function is still larger than the square of the mean count if τ is smaller than τ_c so that both detectors count photons within the same bunch. However, the correlation decreases to become equal to the square of the mean count, that is $g^{(2)} = 1$, for τ much larger than τ_c. Twiss and Little (1959), Morgan and Mandel (1966), and Arecchi et al (1966) have made time delay correlation measurements of this kind, demonstrating the photon bunching effect in a very direct way.

Coherent light on the other hand has no classical intensity fluctuations, or no photon bunches in the quantum picture. The counts in the two detectors are uncorrelated, and only random coincidences occur, giving a correlation function equal to the square of the mean count. The degree of second order coherence is accordingly equal to unity as in equation (5), leading to a zero Hanbury Brown and Twiss result in agreement with the classical prediction in equation (3).

Two Photon Absorption

The statistical or fluctuation properties of a beam of light do not change in a linear absorption experiment, where the rate of absorption of the light is proportional to the beam intensity or photon number. The only effect of the absorption is a reduction of the intensity or photon number distribution of the light by a numerical factor determined by the strength of the absorption. Thus for example in the classical theory of linear absorption of chaotic light, the section of beam shown in figure 1a looks the same after partial absorption, except that the scale of the vertical axis must be appropriately expanded. The degree of second order coherence is unchanged.

On the other hand, a nonlinear optical process generally changes the natures of the light beams involved. The process of two photon absorption from a single beam of light provides a simple illustration. Figure 1 shows a classical calculation by Weber (1971) of the effect of two photon absorption on the intensity fluctuations of chaotic light. Part a shows the time dependence of the intensity before any absorption. Parts b and c show the intensity fluctuations of the same section of beam after its passage through increasing thicknesses of two photon absorbing medium. Now the rate of two photon absorption is proportional to the square of the light intensity; it thus produces a rapid erosion of the peaks in the beam intensity fluctuations, while hardly any absorption occurs in the regions of the intensity troughs. The two photon absorption accordingly has a smoothing effect which is clearly seen in figure 1.

In quantum language, the magnitude of the two photon absorption is determined by the coincidence of pairs of photons at the absorbing atoms. The rate is thus proportional to the correlation function or equivalently to the degree of second order coherence, and it follows from equation (5) that chaotic light is two photon absorbed at twice the rate of absorption of coherent light. The removal of coincident photons by the absorption reduces the degree of second order coherence below its initial value, and figure 3 shows the calculated time dependences of $g^{(2)}$ for light which is initially chaotic and coherent. The rapid fall-off at short times for chaotic light reflects the removal of the coincident pairs associated with the photon bunches. The nature of initially coherent light is also changed as the chance photon coincidences are removed by the two photon absorption. For both kinds of light, the degree of second order coherence ultimately falls to zero when only a single photon is left and no further two photon absorbtion can occur.

Thus for both types of initial light considered, two photon absorption produces light whose degree of second order coherence falls below the unit value characteristic of coherent light. It is seen from equation (8) that such light produces a negative Hanbury Brown and Twiss correlation and it is said to be anticorrelated or antibunched.

The production of antibunched light by two photon absorption of coherent light was predicted theoretically by Chandra and Prakash (1970) and has been studied in detail by Simaan and Loudon (1975) and by Every (1975). Similar effects are produced by any nonlinear process whose rate is enhanced by the occurrence of photon bunches in one of the beams depleted by the process. Another example is second-harmonic generation, recently proposed by Stoler (1974) as a practicable means of producing observable antibunching effects. However, experiments of this kind have not yet been performed.

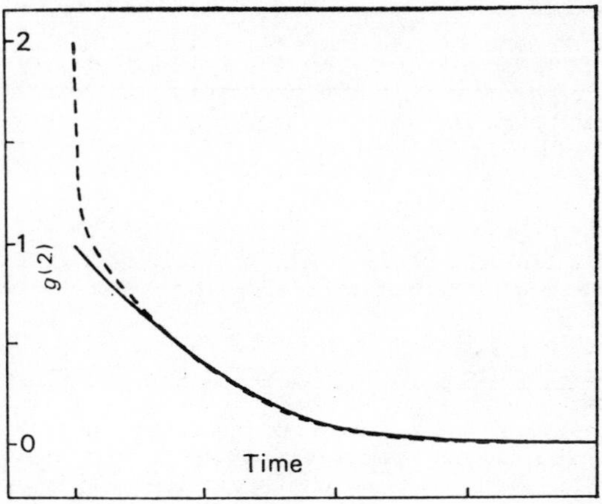

Fig. 3. Effect of two photon absorption on the degree of second order coherence for light which is initially chaotic (broken line) and coherent (solid line) (*after Simaan and Loudon 1975*).

Classical Versus Quantum Theories of Light

Antibunched light has a very important feature for theoretical physics. It cannot be described by a classical theory. This can be seen from the classical definition (4); there is no function of time $\bar{I}(t)$ for the classical fluctuating intensity whose mean-square value can be smaller than the square of its mean value. Thus the minimum classical $g^{(2)}$ is unity, a restriction which allows a classical description of the fluctuation properties of chaotic and coherent light but excludes antibunched light.

There is no similar restriction in the quantum theory of light, and the quantum definition (7) of the degree of second order coherence allows all values from zero to infinity. Table 1 shows some examples of the way in which this definition in terms of a Hanbury-Brown and Twiss experiment can produce small values of $g^{(2)}$. The cases considered have incident light with a definite small number n of photons, and they show in a simple way how the measured degrees of second-order coherence can be smaller than unity (in fact equal to $(n - 1)/n$).

The important feature of the quantum picture is again the limitation on each photon either to transmit through or reflect from the mirror, thus reducing the correlation $\langle n_1 n_2 \rangle$ below the value of the squared mean $\bar{n}^2$ for the cases shown in the table. Classical theory would correspond to splitting each

Table 1. The columns show respectively the number of incident photons in a Hanbury-Brown and Twiss experiment, the possible numbers detected by the two phototubes, their mean, their correlation, and the degree of second order coherence.

n	n_1	n_2	$\bar{n}$	$\langle n_1 n_2 \rangle$	$g^{(2)}$
1	1 0	0 1	1/2	0	0
2	2 1 1 0	0 1 1 2	1	1/2	1/2
3	-	-	3/2	3/2	2/3
4	-	-	2	3	3/4

photon into equal transmitted and reflected parts. A sophisticated version of essentially the same half silvered mirror experiment envisaged in the table has recently been performed (Clauser 1974) with the conclusion that it is the quantum description which fits the observations. It should be emphasized that an analysis similar to that in the table for incident chaotic or coherent light gives the results (5).

The list of experiments whose explanation requires the concept of photons and hence the quantum theory of light is surprisingly short, since a wide range of radiative processes can be described by theories in which the light is treated classically and quantum theory is applied only to atoms involved in the process, the so-called semiclassical approach. The successes of this method have been reviewed by Scully and Sargent (1972); they include for example the photoelectric effect (Lamb and Scully 1969) and stimulated emission, leading to a semiclassical theory of the laser which is adequate for many purposes.

The quantum theory of light merely reproduces the results of semiclassical theory in these cases. Quantum theory is however the more general method and embraces phenomena excluded or covered only approximately by the semiclassical approach, for example spontaneous emission and the Lamb shift.

The photon antibunching effect is a rather direct manifestation of quantum theory and its detailed observation would broaden the range of optical fluctuation phenomena into a novel and intrinsically quantum mechanical domain.

Further Reading

Arecchi F T, Gatti E and Sona A 1966 *Phys. Lett.* *20* 27

Chandra N and Prakash H 1970 *Phys. Rev. A1* 1696

Clauser J F 1974 *Phys. Rev. D9* 853

Cummins H Z and Pike E R (eds) 1974 *Photon Correlation and Light Beating Spectroscopy* (New York: Plenum)

Every I M 1975 *J. Phys. A: Math. Gen.* *8* L69

Hanbury Brown R 1974 *The Intensity Interferometer* (London: Taylor and Francis)

Lamb W E Jr and Scully M O 1969 *Polarization Matière et Rayonnement* ed Société Francaise de Physique (Paris: Presses Universitaires de France) p363

Loudon R 1973 *The Quantum Theory of Light* (Oxford: Clarendon Press)

Morgan B L and Mandel L 1966 *Phys. Rev. Lett.* *16* 1012

Scully M O and Sargent M III 1972 *Physics Today* March *25* 38

Simaan H D and Loudon R 1975 *J. Phys. A: Math. Gen.* *8* 539

Stoler D 1974 *Phys. Rev. Lett.* *33* 1397

Twiss R Q and Little A G 1959 *Aust. J. Phys.* *12* 77

Weber H P 1971 IEEE *J. Quantum Electron.* *7* 189

Reprinted from *Nature*, **280** (1979), 451–454

Evidence for the Quantum Nature of Light

D. F. WALLS

Department of Physics, University of Waikato, Hamilton, New Zealand

A unique property predicted by the quantum theory of light is the phenomenon of photon antibunching. Recent theoretical predictions and experimental observations of photon antibunching in resonance fluorescence from a two-level atom are reviewed.

The advent of photon correlation experiments, as pioneered by Hanbury-Brown and Twiss[1], began a new era in the field of optics. The experiments of classical optics involved a measurement of the first-order correlation function of the electromagnetic field and as such involved only the interference between the probability amplitudes of a single photon. Phenomena such as diffraction, Young's interference experiment and spectral measurements may be categorized as being in the domain of one photon or linear optics. Photon correlation experiments represent a fundamental deviation from these experiments as they involve the interference between different photons and as such are in the realm of nonlinear optics. The quantum-mechanical interpretation of the Hanbury-Brown Twiss effect was given by Glauber[2]. However, adequate explanations of the Hanbury-Brown Twiss effect and related photon correlation experiments have been given which invoke a classical description of a fluctuating electromagnetic field. However, Glauber[2] pointed out that photon correlation experiments offer the possibility of observing a uniquely quantum-mechanical effect - namely photon antibunching. We describe here the recent prediction by Carmichael and Walls[3] and observation by Kimble, Dagenais and Mandel[4-6] and G. Leuchs, M. Rateike and H. Walther (personal communication) of this unique quantum-mechanical effect in photon correlation experiments of fluorescent light from a two-level atom. This marks a dramatic change from all previous photon correlation experiments since the photon antibunching observed cannot be explained on the basis of a classical description of the radiation field. We begin with a brief outline of events leading up to the present investigations.

THE QUANTUM THEORY OF LIGHT

The quantum theory of light beginning with Planck[7] and Einstein[8] played a central part in the development of quantum theory during this century. As the quantum theory was developed a sophisticated theory for the interaction of photons and electrons namely quantum electrodynamics evolved. Amongst the predictions of quantum electrodynamics was that the emission of light from an atom would experience a small shift away from the resonance line of the atom. The experimental observation of this shift, known as the Lamb shift, came as a major triumph for the quantum theory of light[9]. However, for many experiments in optics especially in the field of physical optics the classical picture of light as propagating waves of electromagnetic radiation was adequate to describe the observations. The idea developed that it was not necessary to quantise the light field but only necessary to quantise the atoms and thus describe their interaction by a semiclassical theory. Modifications of this concept such as neoclassical radiation theory also developed. These theories can explain phenomena such as the photoelectric effect, spontaneous emission and even give estimates of the Lamb shift. (A discussion on the present status of such theories as an alternative to quantum electrodynamics is given in ref.10). An early attempt to find quantum effects in a physical optics experiment was made by Taylor[11] who performed Young's interference experiment at very low intensities such that on the average only one photon was incident on the screen at a time. Integrating over a long detection time Taylor obtained an interference pattern as predicted by classical wave theory with no evidence of any unique quantum effects. Taylor's experiment may also be explained on the basis of the quantum theory which interprets it as the interference of the quantum-mechanical probability amplitudes for the photon to go through one slit or the other[12,13]. Essentially the experiments of physical optics were in the regime of one photon or linear optics. We need to go outside the experiments of one photon optics in order to detect any uniquely quantum-mechanical effects.

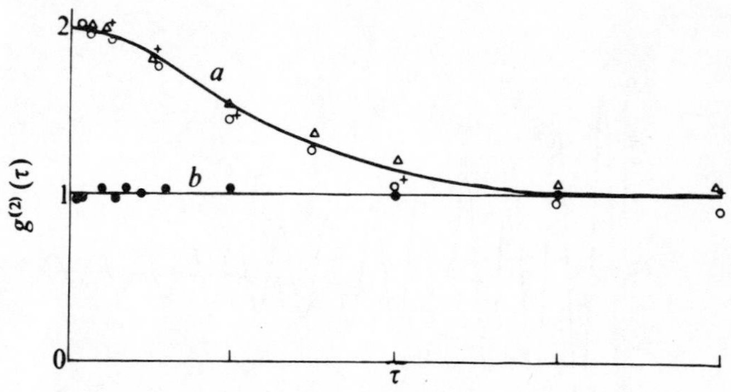

Fig.1 Second order correlation function $g^{(2)}(\tau)$ for: a, thermal light; b, coherent light. Experimental points from Arrechi et al[17].

PHOTON CORRELATION EXPERIMENTS

The first experiment outside the domain of one photon optics was performed by Hanbury-Brown and Twiss in 1956. They performed an intensity correlation experiment, that is a measurement of $\langle I(t)I(t+\tau)\rangle$. Although the original experiment involved the analogue correlation of photocurrents, later experiments used photon counters and digital correlators[14-18]. In essence these experiments measure the joint photocount probability of detecting the arrival of a photon at time t and another photon at time $t + \tau$. The measurements made in this experiment may be described in terms of the second-order correlation function of the radiation field introduced by Glauber[2] in his formulation of optical coherence theory. This is defined as

$$g^{(2)}(\tau) = \frac{\langle E^{(-)}(t)E^{(-)}(t+\tau)E^{(+)}(t+\tau)E^{(+)}(t)\rangle}{(\langle E^{(-)}(t)E^{(+)}(t)\rangle)^2} \tag{1}$$

where $E^{(+)}(t)$ and $E^{(-)}(t)$ are the positive and negative frequency components of the electromagnetic field respectively.

The result of a photon correlation measurement of $g^{(2)}(\tau)$ for a thermal light source using photomultipliers and digital correlators is shown in Fig.1 curve a. We see that the joint counting rate for zero time delay is twice the coincidence rate for large time delays τ. That is, there is a tendency for the photons to arrive in pairs, or a photon-bunching effect. The decay time of the correlations is given by the inverse bandwidth of the light source.

If instead of a thermal light source the experiment is performed with a highly stabilised laser source one obtains a constant correlation function as shown in Fig.1 curve b. This result holds even if the laser and thermal light source have the same bandwidth. Hence there is some fundamental difference between a laser and a thermal light source which may not be apparent in the first-order correlation function but which is manifest in the second-order correlation function. This effect may be understood by considering the photon statistics of different light fields.

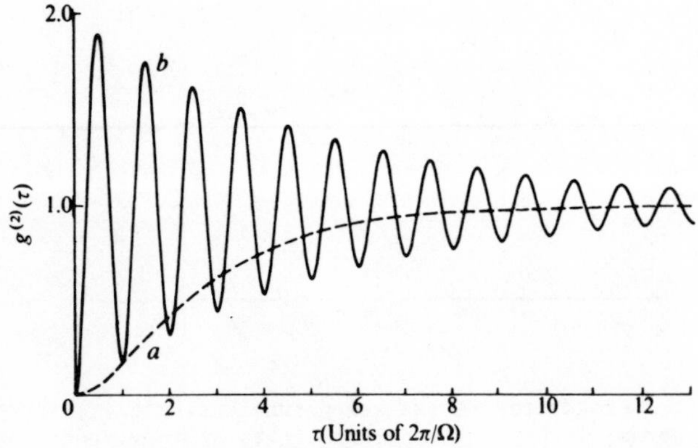

Fig.2 Second order correlation function $g^{(2)}(\tau)$ for fluorescent light from a two level atom: a, low driving field intensity $\Omega \ll \gamma$; b, high driving field intensity $\Omega \gg \gamma$. Theoretical predictions from Carmichael and Walls.[3]

QUANTUM THEORY OF PHOTON CORRELATIONS

We shall attempt in this section to give a quantum theoretic interpretation of photon correlation experiments. For simplicity we shall concentrate on an interpretation of $g^{(2)}(0)$ which enables us to restrict our attention to a single mode field. For a single mode field we may write $g^{(2)}(0)$ in the form

$$g^{(2)}(0) = 1 + \frac{\sigma^2 - \bar{n}}{\bar{n}^2} \tag{2}$$

where $\bar{n}$ and σ^2 are the mean and variance respectively of the photon number distribution. Thermal light has a power law photon number distribution with variance $\sigma^2 = \bar{n}^2 + \bar{n}$ which gives $g^{(2)}(0) = 2$. Coherent light on the other hand as produced by a highly stabilised laser has a poissonian photon number distribution[19-21] with $\sigma^2 = \bar{n}$ leading to $g^{(2)}(0) = 1$. This clearly yields the results shown in Fig.1 for $g^{(2)}(0)$. To include the decay time of the correlation function one must consider many modes. For a field which has a photon number distribution narrower than a Poisson ($\sigma^2 < \bar{n}$) it is possible in principle to obtain a $g^{(2)}(0) < 1$. That is the photon coincidence rate for zero time delay is less than that for large time delays τ. This is the opposite effect to the photon bunching observed for thermal light and has been called photon antibunching. This represents a reduction in the photon number fluctuations below that of a poissonian distribution. The extreme case of photon antibunching would be a light beam where the photons arrive evenly spaced. For a single mode field this is equivalent to having a fixed photon number N which gives $g^{(2)}(0) = 1 - 1/N$.

Let us now consider a classical description of the electromagnetic field. We consider a field described by a fluctuating amplitude E. These fluctuations are taken into account by introducing a probability distribution P(E) for the complex field amplitude. A calculation of $g^{(2)}(0)$ yields

$$g^{(2)}(0) - 1 = \frac{\int P(E)(|E|^2 - \langle|E|^2\rangle)^2 d^2E}{(\langle|E|^2\rangle)^2} \tag{3}$$

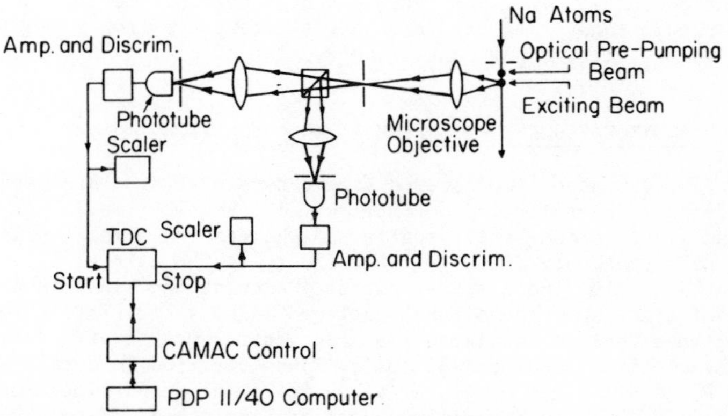

Fig.3 Experimental arrangement used by Kimble et al.[4]

For a thermal field which has a gaussian distribution for P(E) we recover $g^{(2)}(0) = 2$ whereas for a coherent field with a stabilised amplitude P(E) is a delta function and hence $g^{(2)}(0) = 1$. Hence the Hanbury-Brown and Twiss effect for chaotic fields and indeed for the coherent laser field is adequately described by classical theory. However, since the right hand side of equation (3) is a positive semi-definite quantity, we see that classical theory requires $g^{(2)}(0) \geqslant 1$ and hence does not allow photon antibunching.

The formulation of the quantum theory closest in formal appearance to the classical theory involves the diagonal expansion of the density operator of the radiation field in terms of coherent states[2,22],

$$\rho = \int P(\mathcal{E})|\mathcal{E}\rangle\langle\mathcal{E}|\, d^2\mathcal{E} \tag{4}$$

where the coherent states $|\mathcal{E}\rangle$ are eigenstates of the positive frequency part of the electromagnetic field.

$$E^{(+)}|\mathcal{E}\rangle = \mathcal{E}\,|\mathcal{E}\rangle \tag{5}$$

with eigenvalue . The function $P(\mathcal{E})$ does not have the character of a probability distribution since for certain nonclassical states of the radiation field it may be negative or highly singular.

A calculation of $g^{(2)}(0)$ using the P representation yields

$$g^{(2)}(0) - 1 = \frac{\int P(\mathcal{E})(|\mathcal{E}|^2 - \langle|\mathcal{E}|^2\rangle)^2 d^2\mathcal{E}}{\langle|\mathcal{E}|^2\rangle^2} \tag{6}$$

This appears similar in form to the classical expression of equation (3), however, since $P(\mathcal{E})$ may for certain fields which have no classical description take on negative values then $g^{(2)}(0)$ may be less than unity in quantum theory.

While this particular feature of quantum theory was recognised it remained to find a light field which exhibited the property of photon antibunching. It was suggested that certain processes of nonlinear optics for example sub-second harmonic generation[23-26] and two-photon absorption[27-33] would exhibit photon antibunching. To date there has been no experimental verification of these predictions. It was resonance fluorescence from a two-level atom that led to the first experimental observation of photon antibunching.

RESONANCE FLUORESCENCE FROM A TWO-LEVEL ATOM

The topic of resonance fluorescence from a two-level atom has been the subject of considerable theoretical and experimental investigation. For weak incident fields the light is coherently scattered, whereas for strong incident fields when the Rabi frequency Ω ($\Omega = 2\kappa\mathcal{E}/\hbar$ where $\mathcal{E}$ is the driving field amplitude and κ is the atomic dipole matrix element) exceeds the Einstein A coefficient γ of the atom the spectrum of the scattered light splits into three peaks with the two sidebands displaced from the central peak by the Rabi frequency. This spectrum first predicted by Mollow[34] was verified in detail by the experiments of Wu et al.[35] and Hartig et al.[36] following an earlier experiment of Schuda et al.[37] These experiments used an atomic beam of sodium atoms which were optically pumped to prepare a pure two-level system. These atoms were then subjected to irradiation from a highly stabilised dye laser tuned to resonance with the atomic transition. This system was to prove of further interest to physicists.

Carmichael and Walls[3] predicted that the second-order correlation function of the light emitted by a single atom undergoing resonance fluorescence would exhibit the property of photon antibunching. This was confirmed in subsequent calculations by Cohen-Tannoudji[38] and Kimble and Mandel[39]. The results obtained by Carmichael and Walls for the second order correlation function $g^{(2)}(\tau)$ in the steady state are

$$g^{(2)}(\tau) = \left[1 - \exp\left(\frac{-\gamma\tau}{2}\right)\right]^2 \qquad \Omega \ll \gamma \tag{7}$$

$$g^{(2)}(\tau) = \left[1 - \exp\left(\frac{-3\gamma}{4}\tau\right) \cos \Omega\tau\right] \qquad \Omega \gg \gamma \tag{8}$$

in the limiting cases of very weak and very strong driving fields. The behaviour of $g^{(2)}(\tau)$ is plotted in Fig.2. This system clearly displays the property of photon antibunching since in both limits $g^{(2)}(\tau)$ starts at zero. For low-intensity driving fields $g^{(2)}(\tau)$ rises monotonically to a background value of unity, whereas for high driving field intensities $g^{(2)}(\tau)$ rises above unity before reaching a steady state value of unity by damped oscillations.

This behaviour can be understood as follows. A measurement of $g^{(2)}(\tau)$ records the joint probability for the arrival of a photon at time $t = 0$ and the arrival of a photon at time $t = \tau$. Consider now the driven two-level atom as our source of photons. The detection of a fluorescent photon prepares the atom in its ground state since it has just emitted this photon. The probability of seeing a second photon at $\tau = 0$ is zero since the atom cannot re-radiate from the ground state. One must allow some time to elapse so that there may be a finite probability for the atom to be in the excited state and hence a finite probability for the emission of a second photon. In fact $g^{(2)}(\tau)$ is proportional to the probability that the atom will be in its excited state at time τ given that it was initially in the ground state.

Thus a single atom undergoing resonance fluorescence became a candidate for observing photon antibunching. However, the results (equations (7), (8)) hold only for resonance fluorescence from a single atom. If photons from many atoms contribute to the signal detected then one gets interference effects and the antibunching is diminished and for a large number of atoms lost entirely. In fact for a large number of independently contributing atoms one finds $g^{(2)}(\tau) = 2$ in agreement with the central limit theorem[3].

It is clear therefore that an experiment of high sensitivity was necessary if measurements were to be made on the second order correlation function of light from a single atom. Such an experiment was performed by Kimble, Dagenais and Mandel[4]. A sketch of their experimental setup is shown in Fig.3. In a similar configuration to the experiments which measured the spectrum they used an atomic beam of sodium atoms optically pumped to prepare a pure two-level system. The atomic beam was irradiated at right angles with a highly stabilised dye laser tuned on resonance with the $3P_{3/2}$, $F = 3$, $M_F = 2$ to $3^2S_{1/2}$, $F = 2$, $M_F = 2$ transition in sodium. The intensity of the atomic beam was reduced so that on the average no more than one atom is present in the observation region at a time. The fluorescent light from a small observation

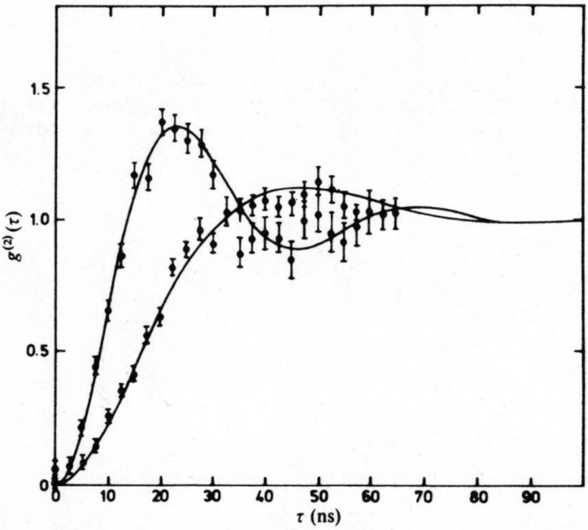

Fig.4 Photon correlation measurements of fluorescent light from sodium. Experimental results obtained by Dagenais and Mandel compared with theory (solid line). ●, Ω/γ = 2.2; O, Ω/γ = 1.1.

volume is observed in a direction orthogonal to both the atomic and laser beams.

The fluorescent light in this direction is divided into two equal parts by a beam splitter and the arrival of photons in each beam is detected by two photomultipliers. The pulses from the two detectors are fed to the start and stop inputs of a time to digital converter (TDC) where the time intervals τ between start and stop pulses are digitised in units of 0.5 ns and stored. The number of events $n(\tau)$ stored at address τ is therefore a measure of the joint photoelectric detection probability density which equals $\eta^2 g^{(2)}(\tau)$ where η is the detector efficiency. The initial results obtained by Kimble et al.[4] showed the initial positive slope of $g^{(2)}(\tau)$ characteristic of photon antibunching but starting with $g^{(2)}(0) = 1$ rather than zero.

The reason for this disagreement with the theory was pointed out by Jakeman et al.[40] who attributed it to number fluctuations in the atomic beam. Though the atomic beam density is such that on the average only one atom is in the observation volume during a correlation time, there are poissonian fluctuations about this mean value so that at times there could be two or more atoms in the observation volume. Thus in the atomic beam experiment one observes the photon antibunching superimposed on the poissonian number fluctuations of the atoms. A calculation including the atomic number fluctuations was carried out by Carmichael et al.[41] Similar calculations were made by Kimble et al.[5] who obtained very good agreement with experimental results. Some recent experimental results of Dagenais and Mandel[6] are shown in Fig.4, where when the effects of atomic number fluctuations are taken into account excellent agreement is obtained with the theory. A photon correlation experiment on resonance fluorescence from a beam of sodium atoms has been recently perfor-

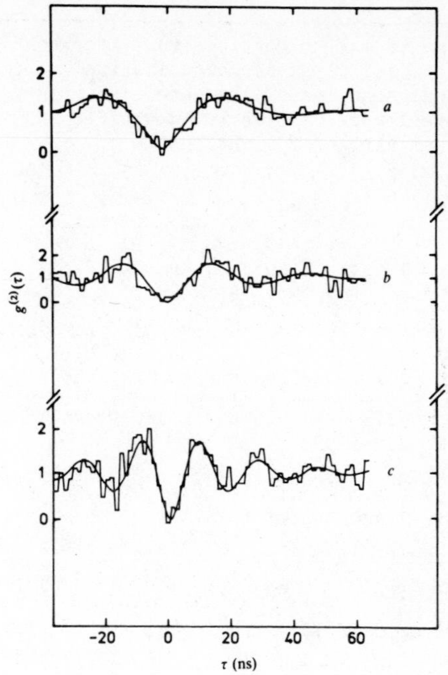

Fig.5 Photon correlation measurements of fluorescent light from sodium. Experimental points adapted from the results of Leuchs et al. (personal communication) compared with theory. a, $\Omega/\gamma = 2.2$; b, $\Omega/\gamma = 4.3$; c, $\Omega/\gamma = 5.7$.

med by Leuchs et al. (personal communication). The results of this experiment are shown in Fig.5, where with compensation for the atomic number fluctuations excellent agreement is found with the theory. These experiments show clear evidence for the existence of photon antibunching and thus verify the predictions of the quantum theory of light.

DISCUSSION

It could still be argued that one has not observed light with the antibunching property but only inferred from the experimental observation that the fluorescent light from a single atom must possess the antibunching property. This is due to the finite correlation at $\tau = 0$ arising from the number fluctuations of the atoms. It would be useful, therefore, to consider ways of improving experimental methods. Recently advances have been made towards isolating a single atom in an ion trap[42]. This suggests the possibility of performing resonance fluorescence on a single stationary atom which should radiate light with the antibunching character. In addition there are the suggested systems in nonlinear optics for example, sub-second harmonic generation[23-26] and two-photon absorption[27-33] where photon antibunching has been predicted. These experiments involve the nonlinear coupling of electromagnetic waves through a medium which may be in the solid state thereby eliminating problems due to atomic number fluctuations.

Although minor refinements may be possible the experiments performed by Kimble et al.[4-6] and Leuchs et al. (personal communication) represent a fundamental contribution to our understanding of the nature of light. For the first time photon correlation experiments have detected an effect which is a direct manifestation of the quantum nature of light.

REFERENCES

1. Hanbury-Brown R and Twiss R Q 1956 *Nature 177* 27; 1957 *Proc. Roy. Soc. 242A* 300; 1957 *Proc. Roy. Soc. 243A* 291.

2. Glauber R J 1963 *Phys. Rev. 130* 2529; 1963 *J. Phys. Rev. 131* 2766; in *Quantum Optics and Electronics*, (eds. de Witt, C. Blandin, A. and Cohen-Tannoudji, C) (Gordon and Breach, Edinburgh, 1964).

3. Carmichael H J and Walls D F 1975 *2nd Natn. Quantum Electronics Conf.;* 1976 *J. Phys. 9B, L43* 1199.

4. Kimble H J, Dagenais M and Mandel L 1977 *Phys. Rev. Lett 39* 691.

5. Kimble H J, Dagenais M and Mandel L 1978 *Phys. Rev. 18A* 201.

6. Dagenais M and Mandel L 1978 *Phys. Rev. 18A* 2217.

7. Planck M 1900 *Vert. dt. Phys. Ges. 2* 202.

8. Einstein A 1905 *Phys. Z 17* 132; 1917 *Phys. Z. 18* 121.

9. Lamb W E 1947 *Phys. Rev. 128* 885.

10. Senitzky I R, Dente G C, Sachs M and Jaynes E T in *Coherence and Quantum Optics*, Vol.4 (eds. Mandel L and Wolf E)(Plenum, New York, 1978).

11. Taylor G I 1909 *Proc. Cam. Phil. Soc. Math. Phys. Sci. 15* 114.

12. Dirac P A M *Quantum Mechanics* (Oxford University Press, 1958).

13. Walls D F 1977 *Am. J. Phys. 45* 952.

14. Rebka G A and Pound R V 1957 *Nature 180* 1035.

15. Twiss R Q and Little G A 1959 *Aust. J. Phys. 12* 77.

16. Morgan B L and Mandel L 1964 *Phys. Rev. Lett. 33* 1397.

17. Arecchi F T, Gatti E and Sona A 1966 *Phys. Lett. 20* 27.

18. Cummins H Z and Pike E R (eds) *Photon Correlation and Light Beating Spectroscopy* (Plenum, New York, 1974).

19. Haken H *Handbuch der Physik* Vol. XXV/2c (Springer, Berlin, 1970).

20. Louisell W H *Quantum Statistical Properties of Radiation* (Wiley, New York, 1973).

21. Sargent M, Scully M O and Lamb W E *Laser Physics* (Addison Wesley, New York, 1974).

22. Sudarshan E C G 1963 *Phys. Rev. Lett. 10* 277.

23. Stoler D 1974 *Phys. Rev. Lett. 33* 1397.

24. Kielich S, Kozierowski M and Tanas R in *Coherence and Quantum Optics IV* (eds Mandel L and Wolf E)(Plenum, New York, 1978).

25. Mostowski J and Rzazewski K 1978 *Phys. Lett. 66A* 275.

26. Drummond P D, McNeil K J and Walls D F 1978 *Opt. Commun.* *28* 255.

27. Chandra N and Prakash H 1970 *Phys Rev. A1*, 1696.

28. Tornau N and Bach A 1974 *Opt. Commun.* *11* 46.

29. Simaan H D and Loudon R 1975 *J. Phys.* *8A* 539.

30. Bandilla A and Ritze H 1976 *Opt. Commun.* *19* 169.

31. Brunner W, Mohr U and Paul H 1976 *Opt. Commun.* *17* 145.

32. Loudon R 1976 *Phys. Bull.* *27* 21.

33. Chaturvedi S, Drummond P and Walls D F 1977 *J Phys.* *10A* L187.

34. Mollow B R 1969 *Phys. Rev.* *188* 1969.

35. Wu F Y, Grove R E and Ezekiel S 1975 *Phys. Rev. Lett.* *35* 1426.

36. Hartig W, Rasmussen W, Schieder R and Walther H 1976 *Z. Phys.* *A278* 205.

37. Schuda F, Stroud C R and Hercher M 1974 *J. Phys.* *B7* L198.

38. Cohen-Tannoudji C in *Frontiers in Laser Spectroscopy* (Eds. Balian R, Haroche S and Liberman S) (North-Holland, Amsterdam, 1977).

39. Kimble H J and Mandel L 1976 *Phys. Rev.* *A13* 2123.

40. Jakeman E, Pike E R, Pusey P N and Vaughan J M 1977 *J. Phys.* *10A*, L257.

41. Carmichael H J, Drummond P, Meystre P and Walls D F 1978 *J. Phys.* *11A* L121.

42. Neuhauser W, Hohenstatt M, Toschek P and Dehmelt H 1978 *Phys. Rev. Lett.* *41* 233.

Reprinted from *Nature*, **177** (1956), 27–29

Correlation between Photons in Two Coherent Beams of Light

R. HANBURY-BROWN and R. Q. TWISS
University of Manchester, Jodrell Bank Experimental Station, and Services Electronics Research Laboratory, Baldock, UK

In an earlier paper[1], we have described a new type of interferometer which has been used to measure the angular diameter of radio stars[2]. In this instrument the signals from two aerials A_1 and A_2 (Fig. 1a) are detected independently and the correlation between the low-frequency outputs of the detectors is recorded. The relative phases of the two radio signals are therefore lost, and only the correlation in their intensity fluctuations is measured; so that the principle differs radically from that of the familiar Michelson interferometer where the signals are combined before detection and where their relative phase must be preserved.

This new system was developed for use with very long base-lines, and experimentally it has proved to be largely free of the effects of ionospheric scintillation[2]. These advantages led us to suggest[1] that the principle might be applied to the measurement of the angular diameter of visual stars. Thus one could replace the two aerials by two mirrors M_1, M_2 (Fig. 1b) and the radio-frequency detectors by photoelectric cells C_1, C_2, and measure, as a function of the separation of the mirrors, the correlation between the fluctuations in the currents from the cells when illuminated by a star.

It is, of course, essential to the operation of such a system that the time of arrival of photons at the two photocathodes should be correlated when the light beams incident upon the two mirrors are coherent. However, so far as we know, this fundamental effect has never been directly observed with light, and indeed its very existence has been questioned. Furthermore, it was by no means certain that the correlation would be fully preserved in the process of photoelectric emission. For these reasons a laboratory experiment was carried out as described below.

The apparatus is shown in outline in Fig. 2. A light source was formed by a small rectangular aperture, 0.13 mm x 0.15 mm in cross-section, on which the image of a high-pressure mercury arc was focused. The 4358 A line was isolated by a system of filters, and the beam was divided by the half-silvered mirror M to illuminate the cathodes of the photomultipliers C_1, C_2. The two cathodes were at a distance of 2.65 m from the source and their areas were limited by identical rectangular apertures O_1, O_2, 9.0 mm x 8.5 mm in cross-section. (It can be shown that for this type of instrument the two cathodes

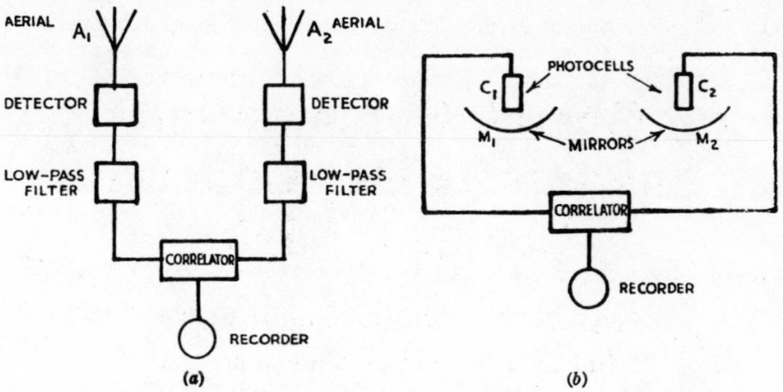

Fig. 1. A new type of radio interferometer (a), together with its analogue (b) at optical wavelengths.

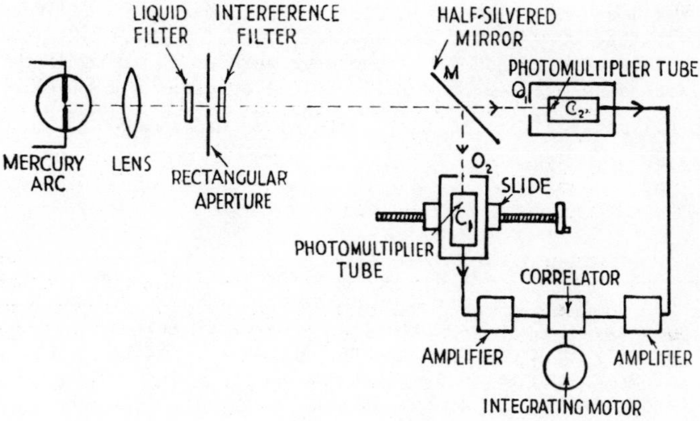

Fig. 2. Simplied diagram of the apparatus.

need not be located at precisely equal distances from the source. In the present case their distances were adjusted to be roughly equal to an accuracy of about 1 cm). In order that the degree of coherence of the two light beams might be varied at will, the photomultiplier C_1 was mounted on a horizontal slide which could be traversed normal to the incident light. The two cathode apertures, as viewed from the source, could thus be superimposed or separated by any amount up to about three times their own width. The fluctuations in the output currents from the photomultipliers were amplified over the band 3-27 Mc/s and multiplied together in a linear mixer. The average value of the product, which was recorded on the revolution counter of an integrating motor, gave a measure of the correlation in the fluctuations. To obtain a significance result it was necessary to integrate for periods of the order of one hour, so very great care had to be taken in the design of the electronic equipment to eliminate the effects of drift, of interference and of amplifier noise.

Assuming that the probability of emission of a photoelectron is proportional to the square of the amplitude of the incident light, one can use classical electromagnetic wave theory to calculate the correlation between the fluctuations in the current from the two cathodes. On this assumption it can be shown that, with the two cathodes superimposed, the correlation S(0) is given by:

$$S(0) = A.T.b_\nu.f\left(\frac{a_1\theta_1\pi}{\lambda_o}\right).f\left(\frac{a_2\theta_2\pi}{\lambda_o}\right)\int\alpha^2(\nu).n_o^{\,2}(\nu).d\nu \qquad (1)$$

It can also be shown that the associated root-mean-square fluctuations N are given by:

$$N = A.T.\frac{2m}{m-1}.\,b_\nu(b_\nu T)^{-\frac{1}{2}}\int\alpha(\nu).n_o(\nu).d\nu \qquad (2)$$

where A is a constant of proportionality depending on the amplifier gain, etc.; T is the time of observation; $\alpha(\nu)$ is the quantum efficiency of the photocathodes at a frequency ν; $n_o(\nu)$ is the number of quanta incident on a photocathode per second, per cycle bandwidth; b_ν is the bandwidth of the amplifiers; $m/(m-1)$ is the familiar excess noise introduced by secondary multiplication; a_1, a_2 are the horizontal and vertical dimensions of the photocathode apertures; θ_1, θ_2 are the angular dimensions of the source as viewed from the photocathodes; and λ_o is the mean wavelength of the light. The integrals are taken over the complete optical spectrum and the phototubes are assumed to be identical. The factor $f(\frac{a\theta\pi}{\lambda_o})$ is determined by the dimensionless parameter η defined by

$$\eta = a\theta/\lambda_o \qquad (3)$$

which is a measure of the degree to which the light is coherent over a photocathode. When $\eta \ll 1$, as for a point source, $f(\eta)$ is effectively unity; however, in the laboratory experiment it proved convenient to make η_1, η_2 of the order of unity in order to increase the light incident on the cathodes and thereby improve the ratio of signal to noise. The corresponding values of $f(\eta_1)$, $f(\eta_2)$ were 0.62 and 0.69 respectively.

When the centres of the cathodes, as viewed from the source, are displaced horizontally by a distance d, the theoretical value of the correlation decreases in a manner dependent upon the dimensionless parameters, η_1 and d/a_1. In the simple case where $\eta_1 \ll 1$, which would apply to an experiment on a visual star, it can be shown that S(d), the correlation as a function of d, is proportional to the square of the Fourier transform of the intensity distribution across the equivalent line source. However, when $\eta \gtrsim 1$, as in the present experiment, the correlation is determined effectively by the apparent overlap of the cathodes and does not depend critically on the actual width of the source. For this reason no attempt was made in the present experiment to measure the apparent angular size of the source.

The initial observations were taken with the photo-cathodes effectively superimposed ($d = 0$) and with varying intensities of illumination. In all cases a positive correlation was observed which completely disappeared, as expected, when the separation of the photo-cathodes was large. In these first experiments the quantum efficiency of the photo-cathodes was too low to give a satisfactory ratio of signal to noise. However, when an improved type of photomultiplier became available with an appreciably higher quantum efficiency, it was possible to make a quantitative test of the theory.

A set of four runs, each of 90 minutes duration, was made with the cathodes superimposed ($d = 0$), the counter readings being recorded at 5-minute intervals. From these readings an estimate was made of N_e, the root mean square deviation in the final reading $S(0)$ of the counter, and the observed values of $S_e(0)/N_e$ are shown in column 2 of Table 1. The results are given as a ratio in order to eliminate the factor A in equations (1) and (2), which is affected by changes in the gain of the equipment. For each run the factor

$$\frac{m-1}{m}\int \alpha^2(\nu)n_o^{\ 2}(\nu)d\nu \Big/ \int \alpha(\nu)n_o(\nu)d\nu$$

was determined from measurements of the spectrum of the incident light and of the d.c. current, gain and output noise of the photomultipliers; the corres-poinding theoretical values of $S(0)/N$ are shown in the second column of Table 1. In a typical case, the photomultiplier gain was 3×10^5, the output current was 140 μamp., the quantum efficiency $\alpha(\nu_o)$ was of the order of 15 per cent and $n_o(\nu_o)$ was of the order of 3×10^{-3}. After each run a comparison run was taken with the centres of the photo-cathodes, as viewed from the source, separated by twice their width ($d = 2a$), in which position the theoretical correlation is virtually zero. The ratio of $S_e(d)$, the counter reading after 90 minutes, to N_e, the root mean square deviation, is shown in the third column of Table 1.

Table 1. Comparison between the theoretical and experimental values of the correlation

	Cathodes superimposed ($d = 0$)		Cathodes separated ($d = 2a = 1.8$ cm.)	
	Experimental ratio of correlation to r.m.s. deviation $S_e(0)/N_e$	Theoretical ratio of correlation to r.m.s. deviation $S(0)/N$	Experimental ratio of correlation to r.m.s. deviation $S_e(d)/N_e$	Theoretical ratio of correlation to r.m.s. deviation $S(d)/N$
1	+ 7.4	+ 8.4	- 0.4	~ 0
2	+ 6.6	+ 8.0	+ 0.5	~ 0
3	+ 7.6	+ 8.4	+ 1.7	~ 0
4	+ 4.2	+ 5.2	- 0.3	~ 0

The results shown in Table 1 confirm that correlation is observed when the cathodes are superimposed but not when they are widely separated. However, it may be noted that the correlations observed with $d = 0$ are consistently lower than those predicted theoretically. The discrepancy may not be significant but, if it is real, it was possibly caused by defects in the optical system. In particular, the image of the arc showed striations due to imperfections in the glass bulb of the lamp; this implies that unwanted differential phase-shifts were being introduced which would tend to reduce the observed correlation.

This experiment shows beyond question that the photons in two coherent beams of light are correlated, and that this correlation is preserved in the process of photoelectric emission. Furthermore, the quantitative results are in fair agreement with those predicted by classical electromagnetic wave theory and the correspondence principle. It follows that the fundamental principle of the interferometer represented in Fig. 1b is sound, and it is proposed to examine in further detail its application to visual astronomy. The basic mathematical theory together with a description of the electronic apparatus used in the laboratory experiment will be given later.

We thank the Director of Jodrell Bank for making available the necessary facilities, the Superintendent of the Services Electronics Research Laboratory for the loan of equipment, and Mr. J. Rodda, of the Ediswan Company for the use of two experimental photo-tubes. One of us wishes to thank the Admiralty for permission to submit this communication for publication.

REFERENCES

Hanbury Brown R and Twiss R Q 1954 *Phil. Mag.* *45* 663

Jennison R C and Das Gupta M K *Phil. Mag. (8th Series)* 1 55; ibid. 1 65.

Reprinted from *Nature*, **178** (1956), 1449–1450

The Question of Correlation between Photons in Coherent Light Rays

E. M. PURCELL

Lyman Laboratory of Physics, Harvard University, Cambridge, Massachusetts, USA

Brannen and Ferguson[1] have reported experimental results which they believe to be incompatible with the observation by Hanbury-Brown and Twiss[2] of correlation in the fluctuations of two photoelectric currents evoked by coherent beams of light. Brannen and Ferguson suggest that the existence of such a correlation would call for a revision of quantum theory. It is the purpose of this communication to show that the results of the two investigations are not in conflict, the upper limit set by Brannen and Ferguson being in fact vastly greater than the effect to be expected under the conditions of their experiment. Moreover, the Brown-Twiss effect, far from requiring a revision of quantum mechanics, is an instructive illustration of its elementary principles. There is nothing in the argument below that is not implicit in the discussion of Brown and Twiss, but perhaps I may clarify matters by taking a different approach.

Consider first an experiment which is simpler in concept than either of those that have been performed, but which contains the essence of the problem. Let *one* beam of light fall on *one* photomultiplier, and examine the statistical fluctuations in the counting rate. Let the source be nearly monochromatic and arrange the optics so that, as in the experiments already mentioned, the difference in the length of the two light-paths from a point A in the photocathode to two points B and C in the source remains constant, to within a small fraction of a wavelength, as A is moved over the photocathode surface. (This difference need not be small, nor need the path-lengths themselves remain constant.) Now it will be found, even with the steadiest source possible, that the fluctuations in the counting-rate are slightly greater than one would expect in a random sequence of independent events occurring at the same average rate. There is a tendency for the counts to 'clump'. From the quantum point of view this is not surprising. It is typical of fluctuations in a system of bosons. I shall show presently that this extra fluctuation in the single-channel rate necessarily implies the cross-correlation found by Brown and Twiss. But first I propose to examine its origin and calculate its magnitude.

Let P denote the square of the electric field in the light at the cathode surface in one polarization, averaged over a few cycles. P is substantially constant over the photocathode at any instant, but as time goes on it fluctuates in a manner determined by the spectrum of the disturbance, that is, by the 'line shape'. Supposing that the light contains frequencies around ν_0, we describe the line shape by the normalized spectral density $g(\nu - \nu_0)$. The width of the distribution g, whether it be set by circumstances in the source itself or by a filter, determines the rate at which P fluctuates. For our purpose, the stochastic behaviour of P can be described by the correlation function $\overline{P(t)\ P(t+\tau)}$, which is related in turn to $g(\nu - \nu_0)$ by[3]

$$\overline{P(t)\ P(t+\tau)} = \bar{P}^2(1 + |\rho|^2),$$

$$\text{where } \rho = \int_{-\infty}^{\infty} g(x) \exp 2\pi i\tau x \, dx \tag{1}$$

For the probability that a photoelectron will be ejected in time dt, we must write $\alpha P dt$, where α is constant throughout the experiment. It makes no difference whether we think of P as the square of an electric field-strength or as a photon probability density. (In this connection the experiment of Forrester, Gudmundsen and Johnson[4] on the photoelectric mixing of incoherent light is interesting.) Assuming one polarization only, and one count for every photoelectron, we look at the number of counts n_T in a fixed interval T, and at the fluctuations in n_T over a sequence of such intervals. From the above relations, the following is readily derived:

$$\overline{n_T^2} - \bar{n}_T^2 = \bar{n}_T(1 + \alpha\bar{P}\tau_0) \tag{2}$$

$$\text{where } \tau_0 = \int_{-\infty}^{\infty} |\rho|^2 d\tau$$

and it has been assumed in deriving (2) that $T >> \tau_0$. Now $\alpha\bar{P}$ is just the average counting-rate and τ_0, a correlation time determined by the light spectrum, is approximately the reciprocal of the spectral bandwidth $\Delta\nu$; in particular, if $\Delta\nu$ is the full width at half intensity of a Lorentzian density function, $\tau_0 = (\pi\Delta\nu)^{-1}$, while if $\Delta\nu$ is the width of a rectangular density function, $\tau_0 = \Delta\nu^{-1}$. We see that the fractional increase in mean-square fluctuation over the 'normal' amount is independent of T, and is about equal to the number of counts expected in an interval $1/\Delta\nu$. This number will ordinarily be very much smaller than one. The result, expressed in this way, does not depend on the counting efficiency.

If one insists on representing photons by wave packets and demands an explanation in those terms of the extra fluctuation, such an explanation can be given. But I shall have to use language which ought, as a rule, to be used warily. Think, then, of a stream of wave packets, each about $c/\Delta\nu$ long, in a random sequence. There is a certain probability that two such trains accidentally overlap. When this occurs they interfere and one may find (to speak rather loosely) four photons, or none, or something in between as a result. It is proper to speak of interference in this situation because the conditions of the experiment are just such as will ensure that these photons are in the same quantum state. To such interference one may ascribe the 'abnormal' density fluctuations in any assemblage of bosons.

Were we to carry out a similar experiment with a beam of electrons, we should, of course, find a slight suppression of the normal fluctuations instead of a slight enhancement; the accidentally overlapping wave trains are precisely the configurations excluded by the Pauli principle. Nor would we be entitled in that case to treat the wave function as a classical field.

Turning now to the split-beam experiment, let n_1 be the number of counts of one photomultiplier in an interval T, and let n_2 be the number of counts in the other in the same interval. As regards the fluctuations in n_1 alone, from interval to interval, we face the situation already analysed, except that we shall now assume both polarizations present. The fluctuations in orthogonal polarizations are independent, and we have, instead of (2),

$$\overline{\Delta n_1^{\,2}} = \overline{n_1^{\,2}} - \bar{n}_1^{\,2} = \bar{n}_1(1 + \tfrac{1}{2}\bar{n}_1\tau_0/T) \qquad (3)$$

where n_1/T has been written for the average counting-rate in channel 1. A similar relation holds for n_2. Now if we should connect the two photomultiplier outputs together, we would clearly revert to a single-channel experiment with a count $n = n_1 + n_2$. We must then find:

$$\overline{\Delta n^2} = \bar{n}\,(1 + \tfrac{1}{2}\bar{n}\tau_0/T) \qquad (4)$$

But $$\overline{\Delta n^2} = \overline{(\Delta n_1 + \Delta n_2)^2}.$$

$$= \bar{n}_1(1 + \tfrac{1}{2}\bar{n}_1\tau_0/T) + \bar{n}_2(1 + \tfrac{1}{2}\bar{n}_2\tau_0/T) + \overline{2\Delta n_1\Delta n_2} \qquad (5)$$

From (4) and (5) it follows that:

$$\overline{\Delta n_1\Delta n_2} = \tfrac{1}{2}\bar{n}_1^{\,2}\tau_0/T \qquad (6)$$

This is the positive cross-correlation effect of Brown and Twiss, although they express it in a slightly different way. It is merely another consequence of the 'clumping' of the photons. Note that if we had separated the branches by a polarizing filter, rather than a half-silvered mirror, the factor 1/2 would be lacking in (4), and (5) would have led to $\overline{\Delta n_1\Delta n_2} = 0$, which is as it should be.

If we were to split a beam of electrons by a non-polarizing mirror, allowing the beams to fall on separate electron multipliers, the outputs of the latter would show a negative cross-correlation. A split beam of classical particles would, of course, show zero cross-correlation. As usual in fluctuation phenomena, the behaviour of fermions and the behaviour of bosons deviate in opposite directions from that of classical particles. The Brown-Twiss effect is thus, from a *particle* point of view, a characteristic quantum effect.

It remains to show why Brannen and Ferguson did not find the effect. They looked for an increase in coincidence-rate over the 'normal' accidental rate, the latter being established by inserting a delay in one channel. Their single-channel rate was 5 x 10^4 counts per second, their accidental coincidence rate about 20 per second, and their resolving time about 10^{-8} sec. To analyse their experiment one may conveniently take the duration T of an interval of observation to be equal to the resolving time. One then finds that the coincidence-rate should be enhanced, in consequence of the cross-correlation, by the factor $(1 + \tau_0/2T)$. Unfortunately, Brannen and Ferguson do not specify their optical bandwidth; but it seems unlikely, judging from their description of their source, that it was much less than 10^{11} cycles/sec., which corresponds to a spread in wavelength of rather less than 1 A. at 4358 A. Adopting this figure for illustration, we have $\tau_0 = 10^{-11}$ sec., so that the expected fractional change in coincidence-rate is 0.0005. This is much less than the statistical uncertainty in the coincidence-rate in the Brannen and Ferguson experiment, which was about 0.01. Brown and Twiss did not count individual photoelectrons and coincidences, and were able to work with a primary photoelectric current some 10^4 times greater than that of Brannen and Ferguson. It ought to be possible to detect the correlation effect by the method of Brannen and Ferguson. Setting counting efficiency aside, the observing time required is proportional to the resolving time and inversely proportional to the square of the light flux per unit optical bandwidth. Without a substantial increase in the latter quantity, counting periods of the order of years would be needed to demonstrate the effect with the apparatus of Brannen and Ferguson. This only adds lustre to the notable achievement of Brown and Twiss.

REFERENCES

1. Brannen E and Ferguson H I S 1956 *Nature 178* 481
2. Brown H R and Twiss R Q 1956 *Nature 177* 27
3. Lawson J L and Uhlenbeck G E *Threshold Signals*, p.61 (McGraw-Hill, New York, 1950)
4. Forrester A I, Gudmundsen R A and Johnson P O 1955 *Phys. Rev. 99* 1691

Reprinted from *Phys. Rev. Letts.*, **39** (1977), 691–695

Photon Antibunching in Resonance Fluorescence

H. J. KIMBLE*, M. DAGENAIS and
L. MANDEL

Department of Physics and Astronomy,
University of Rochester, Rochester, NY 14627, USA

The phenomenon of antibunching of photoelectric counts has been observed in resonance fluorescence experiments in which sodium atoms are continuously excited by a dye-laser beam. It is pointed out that, unlike photoelectric bunching, which can be given a semiclassical interpretation, antibunching is understandable only in terms of a quantized electromagnetic field. The measurement also provides rather direct evidence for an atom undergoing a quantum jump.

The tendency of photons in a light beam emitted by a thermal equilibrium source to arrive in bunches, rather than strictly at random, has been well known since the classic experiments of Hanbury Brown and Twiss[1]. The bunching phenomenon was studied more explicitly in time-resolved correlation experiments,[2] and it was confirmed that the joint probability density of photo-detection $P_2(t,t+\tau)$ by a phototube at two times t and $t+\tau$ is greatest when τ is near zero, and falls to a constant lower value once τ appreciably exceeds the coherence time. It is possible to look on the bunching phenomenon as a characteristic quantum feature of thermal bosons. If the wave functions of neighbouring bosons overlap, when the states are symmetrized the resulting interference increases the probability of detecting one boson near another one.

Nevertheless, it is well known that we may also account for the bunching of photoelectric pulses in terms of the fluctuations of a completely classical field of instantaneous intensity $I(t) \equiv E_i^*(t)E_i(t)$. The joint probability density $P_2(t,t+\tau)$ can be shown to be given by[3]

$$P_2(t,t+\tau) = \alpha^2 \langle I(t)I(t+\tau)\rangle, \tag{1}$$

where α is a constant characterizing the efficiency of the detector, and the average is to be taken over the ensemble of all realizations of the electro-

---

*Present address: General Motors Research Laboratories, Warren, Michigan 48063, U.S.A.

magnetic field $E_i(t)$. If we introduce the normalized correlation function ($\Delta I \equiv I - \langle I \rangle$)

$$\lambda(\tau) \equiv \langle \Delta I(t) \Delta I(t+\tau) \rangle / \langle I(t) \rangle \langle I(t+\tau) \rangle, \tag{2}$$

we may reexpress Eq. (1) in the form

$$P_2(t,t+\tau) = \alpha^2 \langle I(t) \rangle \langle I(t+\tau) \rangle \, [1 + \lambda(\tau)]. \tag{3}$$

Since $\lambda(\tau) \leq \lambda(0)$ and $\lambda(\infty) = 0$ for an ergodic process, this equation shows at once that the twofold detection probability is greater for time intervals τ near zero than for long intervals.

In the fully quantized treatment of the same problem given by Glauber,[4] $\hat{I}(t)$ becomes a Hilbert-space operator[5] $\hat{E}_i^\dagger(t)\hat{E}_i(t)$, and the correlations have to be written in normal order (::) and in time order (T) in the form $\langle T{:}\hat{I}(t)\hat{I}(t+\tau){:}\rangle$. With this change Eqs. (1) - (3) remain valid. The correlation between the two expressions is provided by the diagonal coherent-state ($|\{v\}\rangle$) representation of the density operator $\hat{\rho}$ of the free field[6,4]:

$$\hat{\rho} = \int \phi(\{v\}) \, |\{v\}\rangle \langle \{v\}| \, d\{v\}. \tag{4}$$

This allows expectation values of normally ordered operators to be expressed just like c-number averages, with the weighting or phase-space functional $\phi(\{v\})$ playing the formal role of the "probability functional." For a quantum field we then merely replace classical averages like $\langle I(t)I(t+\tau) \rangle$ by $\langle I(t)I(t+\tau) \rangle_\phi$. However $\phi(\{v\})$ has the full character of a classical probability functional only for those states of the electromagnetic field for which a classical analogue exists [e.g. for thermal fields having Gaussian $\phi(\{v\})$], when the quantum and classical descriptions lead to similar conclusions.

On the other hand, there also exist quantum states of the field that have no classical description, for which $\phi(\{v\})$ may be negative or highly singular and does not have the character of a probability density. For such states $\lambda(0)$ may become negative, and Eq. (3) then predicts the opposite effect, or antibunching near $\tau = 0$, and the joint probability density $P_2(t,t+\tau)$ may increase with τ from $\tau = 0$. Such antibunching or negative photoelectric correlation is an explicit feature of a quantum field,[7-11] and its observation would provide rather direct evidence for the existence of optical photons, unlike positive correlation effects that have a semiclassical explanation.

It has recently been predicted that the electromagnetic field radiated by a driven two-level atom in the presence of a continuous exciting field has the correlation function[10]

$$\langle T{:}\hat{I}(t)\hat{I}(t+\tau){:}\rangle = \langle \hat{I}(t) \rangle \langle \hat{I}_G(\tau) \rangle, \tag{5}$$

where $\langle \hat{I}_G(\tau) \rangle$ is the mean light intensity radiated at time τ following the turn-on of the interaction when the atom starts in the lower (or ground) state. Since $\langle \hat{I}_G(\tau) \rangle$ is necessarily zero for $\tau = 0$ and increases with τ from zero, such a driven atom is an example of a source exhibiting photon antibunching, with $\lambda(0) = -1$ in the steady state. We wish to report the observation of photoelectric antibunching in resonance fluorescence experiments on single atoms.

The principle of the experiment is illustrated in Fig. 1. Atoms of sodium in an atomic beam are excited by the light beam of a tunable dye laser propagating into the plane of the diagram. The light is stabilized both in intensity (to a few percent) and in frequency (to about 1 MHz). The laser is tuned to the ($3^2S_{1/2}$, F = 2) to ($3^2P_{3/2}$, F = 3) transition, and by optical pre-pumping with circularly polarized light in a weak magnetic field just before the final interaction, the sodium atoms are prepared in the $3^2S_{1/2}$, F = 2, $m_F = 2$ state, from which the only allowed transition is to the $3^2P_{3/2}$, F = 3, $m_F = 3$ state[12]. This procedure assures that the atoms behave like two-level quantum systems. The resonant optical field seen by each atom has a power density of about 70 mW/cm^2, corresponding to a Rabi frequency Ω between 2 and 2.5 times the Einstein A coefficient (in the language of Ref. 10, $\Omega/\beta \sim 4$-5). The fluorescent light emitted is collected at right angles to both the laser and atomic beams by a microscope objective that images a region of linear dimensions of order 100 μm onto an aperture. The light emerging from this aperture is divided into approximately equal parts by a beam splitter, and is then further imaged onto two photomultiplier tubes. The atomic beam also is passed through a 100-μm aperture, and the atomic current is held low enough to ensure that no more than one or two atoms in the field are able to contribute to the collected fluorescence radiation at the same time. After amplification and pulse shaping, the pulses from the two detectors are fed to the start and stop inputs of a time-to-digital converter (TDC), where the time intervals τ between start and stop pulses are digitized in units of 0.5 nsec and stored. The number of events $n(\tau)$ stored at address τ is therefore a measure of the joint photoelectric detection probability density $P_2(t,t+\tau)$. The TDC is under the control of a PDP 11/40 computer, where the information is ultimately recorded.

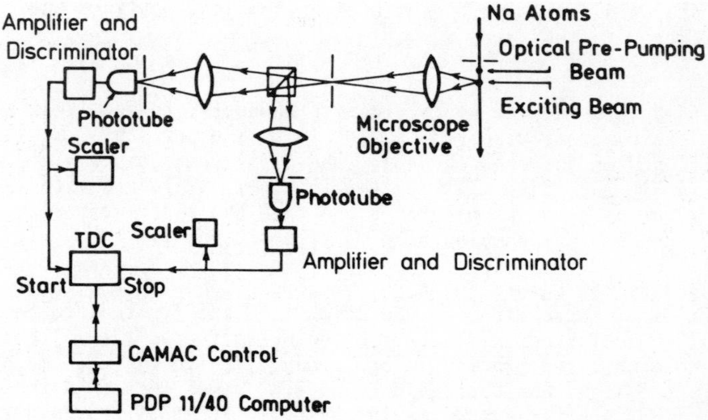

Fig. 1. Outline of the principal elements of the experiment.

Figure 2 shows the number of photoelectric pulse pairs $n(\tau)$ recorded for various time delays τ in intervals of 2 nsec, before any corrections are applied to the data. The number $n(\tau)$ is related to the intensity correlation function at the two detectors by

$$\langle n(\tau)\rangle = N_s \Delta\tau \alpha_2 \langle T:\hat{I}_1(t)\hat{I}_2(t+\tau):\rangle / \langle \hat{I}_1(t)\rangle, \quad (6)$$

where $\hat{I}_1(t)$ and $\hat{I}_2(t)$ are expressed in units of photons per second and suf-

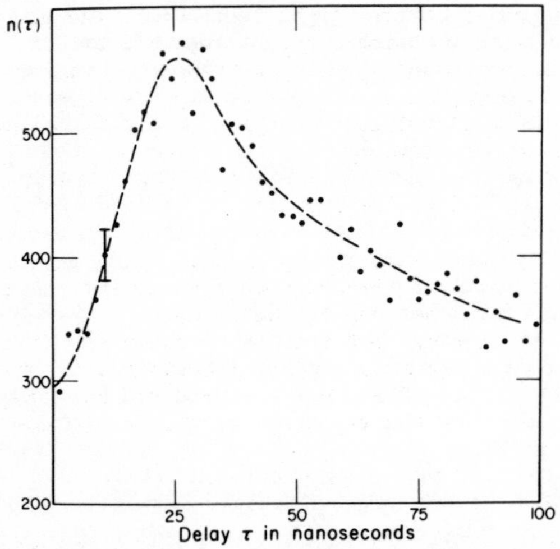

Fig. 2. The number of recorded pulse pairs $n(\tau)$ as a function of the time delay τ in nanoseconds. The growth of $n(\tau)$ from $\tau = 0$ shows antibunching. The bars on one point indicate statistical uncertainties corresponding to one standard deviation. The broken line just outlines the trend.

fixes 1 and 2 refer to the start and stop channels, respectively; N_s is the number of start pulses received (9×10^6 in this experiment); α_2 is the detection efficiency in the stop channel; and $\Delta\tau$ is the channel width (2 nsec for the data in Fig. 2). It will be seen that $n(\tau)$ increases with τ from its smallest value at $\tau = 0$, as predicted, so that the experiment shows unmistakable evidence for antibunching of photoelectric pulses.

In order to attempt a more quantitative comparison with the theory we have to make a number of corrections to the data. In the first place there are accidental pair correlations $n(\tau)$ contributed by scattered laser light that enters the two phototubes and provides a background. Although the amount of scattered background light has been held to several times less than the fluorescent light by use of apertures and imaging systems, it still represents a problem in this experiment, because of the requirement to study single atoms so far as possible. In order to correct for this we express the total field at each detector as the sum of two parts $\hat{E}_i(1) + e_i(1)$ and $\hat{E}_i(2) + e_i(2)$, where $\hat{E}_i$ represents the fluorescent field and e_i the scattered laser field that we treat as a c-number, and we substitute in Eq. (6). The correlation function then breaks into sixteen distinct terms, most of which have been calculated in Ref. 10. If we discard terms in e, e^2, and $|e|^2e$, on the ground that the scattered laser field probably has a rapidly oscillating phase across the cathode of the photodetector, so that the corresponding interference terms sum almost to zero, we find that

$$\langle n(\tau)\rangle \approx (N_s\Delta\tau/R^T{}_1)[R^T{}_1R^T{}_2 + R_1R_2\lambda(\tau) + 2(R_1R_2r_1r_2)^{\frac{1}{2}}|\gamma(\tau)|]. \quad (7)$$

Here R_i, r_i and R^T_i $(i=1,2)$ are the counting rates at photodetector i contributed by the fluorescent light alone, by the scattered laser light alone, and by both fluorescent and scattered light together. The last two are of course measurable directly with the atomic beam turned off and on, and R_i is given by $R^T_i - r_i$. The following counting rates were used in the measurement: R^T_1 = 13 600/sec, R^T_2 = 16 600/sec, r_1 = 2430/sec, r_2 = 4450/sec. $\gamma(\tau)$ is the normalized second-order correlation function of the fluorescent field, which was first calculated by Mollow.[13]

The result of applying Eq. (7) to the data is shown in Fig. 3, where we have plotted $1+\lambda(\tau)$ against τ. It will be seen that the initial value $\lambda(0)$ is approximately -0.6, rather than -1 as predicted theoretically. We believe that this is probably because the detected fluorescence is not always produced by a single atom, but sometimes by two or more. The atomic flux is difficult to measure with accuracy, but from the properties of the detectors and the geometry we estimate the mean number of contributing atoms to be around 1, within a factor of 2 or 3. For the field produced by two radiating atoms located at random positions the expected value of $\lambda(0)$ would be -0.5.

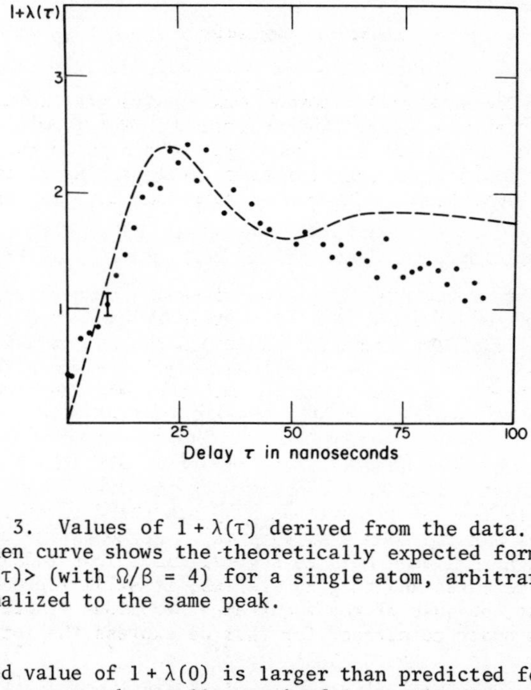

Fig. 3. Values of $1+\lambda(\tau)$ derived from the data. The broken curve shows the theoretically expected form of $<\hat{I}_G(\tau)>$ (with $\Omega/\beta = 4$) for a single atom, arbitrarily normalized to the same peak.

As the derived value of $1+\lambda(0)$ is larger than predicted for a single radiating atom, we cannot make really meaningful quantitative comparisons with the single-atom theory. Nevertheless, in order to show that the measured behaviour of $1+\lambda(\tau)$ has at least the general form predicted theoretically[9,10] for $<\hat{I}_G(\tau)>$, we have also indicated the predicted form of $<\hat{I}_G(\tau)>$ in Fig. 3 for a Rabi frequency $\Omega = 4\beta$, normalized rather arbitrarily to the same maximum. The differences between this curve and the derived values of $1+\lambda(\tau)$ for large τ are partly attributable to the short transit time of atoms moving with average velocity around 10^5 cm/sec through the field of view sampled by the photodetectors. This causes the measured values of

$1 + \lambda(\tau)$ to be biased downwards for large τ as compared with those for short τ. Nevertheless, the measured correlation function is at least qualitatively as expected.

Finally, we would like to point out that the evidence provided for photon antibunching near $\tau = 0$ is, at the same time, rather direct evidence for an atom undergoing a quantum jump. Although the state of an excited atom evolves continuously in the absence of an observation, the theory predicts a sudden return to the lower state when a photon is detected. That $P_2(t,t+\tau)$ vanishes when $\tau = 0$ for a single atom may be regarded as a reflection of the fact that the atom, having emitted a photon at time t, is unable to radiate again immediately after having made a quantum jump back to the lower state. The quantum nature of the radiation field and the quantum jump in emission, which are of course inextricably connected, are therefore both manifest in these photoelectric correlation measurements.

This work was supported by the National Science Foundation.

REFERENCES

1. Hanbury Brown R and Twiss R Q 1956 *Nature (London) 177* 27; 1957 *Proc. Roy. Soc. London, Ser. A 242* 300; and 1957 *243* 291.
2. Morgan B L and Mandel L 1966 *Phys. Rev. Lett. 16* 1012; Scarl D B 1966 *Phys. Rev. Lett. 17* 663; Phillips D T, Kleinman H and Davis S P 1967 *Phys. Rev. 153* 113.
3. Purcell E M 1956 *Nature (London) 178* 1449; Mandel L 1958 *Proc. Phys. Soc., London 72* 1037, and 1959 *74* 233; Mandel L in *Progress in Optics*, edited by Wolf E (North-Holland, Amsterdam, 1963), Vol.2, p.181; Mandel L, Sudarshan E C G and Wolf E 1964 *Proc. Phys. Soc., London 84* 435.
4. Glauber R J 1963 *Phys. Rev. 130* 2529, and 1963 *131* 2766.
5. We label all Hilbert-space operators by the caret.
6. Sudarshan E C G 1963 *Phys. Rev. Lett. 10* 277; Mehta C L and Sudarshan E C G 1965 *Phys. Rev. 138* B274; Klauder J R 1966 *Phys. Rev. Lett. 16* 534; Klauder J R and Sudarshan E C G, *Fundamentals of Quantum Optics* (Benjamin, New York, 1968).
7. We should emphasize that for a classical field $\lambda(\tau)$ is necessarily non-negative only for $\tau = 0$, as is clear from the definition (2). For other values of τ, $\lambda(\tau)$ may be negative even for a classical field; this is a characteristic feature of light beats.
8. Stoler D 1974 *Phys. Rev. Lett. 33* 1397; Kozierowski M and Tanaś R 1977 *Opt. Comm. 21* 229.
9. Carmichael H J and Walls D F 1976 *J Phys. B 9* 1199.
10. Kimble H J and Mandel L 1976 *Phys. Rev. A 13* 2123; and 1977 *15* 689.
11. Mandel L in *Progress in Optics*, edited by Wolf E (North-Holland, Amsterdam, 1976), Vol. 13, p.27, and to be published.
12. Abate J A 1974 *Opt. Comm. 10* 269.
13. Mollow B R 1969 *Phys. Rev. 188* 1969.

Reprinted from *Am. J. Phys.*, **45** (1977), 952–956

A Simple Field Theoretic Description of Photon Interference

D. F. WALLS

Physics Department, School of Science, University of Waikato, Hamilton, New Zealand

A description of photon-interference experiments of the Young's type using the techniques of quantum-field theory is presented. In particular, an explicit illustration of Dirac's statement that a photon interferes only with itself is given. The interference patterns produced by a one photon, an n photon and a coherent photon field are calculated. A discussion of the degree of optical coherence possessed by the above fields is also incorporated.

1. INTRODUCTION

Photon-interference experiments of the kind typified by Young's interference experiment and Michelson's interferometer played a central role in early discussions of the dual wave and corpuscular nature of light. These experiments basically detect the interference pattern resulting from the superposition of two components of a light beam. Classical theory based on the wave nature of light readily yielded the observed interference pattern. For full details of the classical theory and experimental arrangements of these interference experiments the reader is referred to the classic text of Born and Wolf.[1]

An explanation for the interference pattern based on the quantum nature of light was first put forward by Dirac in his classic text on quantum mechanics[2]. Here he argued that the observed intensity pattern results from interference between the probability amplitudes of a single photon to take either of the two possible paths. The crux of the quantum-mechanical explanation is that the wave function gives information about the probability of one photon being in a particular place and not the probable number of photons in that place. Dirac points out that the interference between the two beams does not arise because photons of one beam sometimes annihilate photons from the other and sometimes combine to produce four photons.

"This would contradict the conservation of energy. The new theory which connects the wave functions with probabilities for one photon gets over the difficulty by making each photon go partly into each of the two components. Each photon then interferes only with itself. Interference between two different photons never occurs." We stress that the above quoted statement of Dirac was only intended to apply to experiments of the Young's type where the interference pattern is revealed by detecting single photons.

It is the aim of this paper to explicitly illustrate the above remarks using the techniques of quantum electrodynamics. In particular, we consider a field containing only one photon and show how an interference pattern may emerge from a succession of one-photon-interference experiments.

In order to apply the techniques of quantum electrodynamics to Young's interference experiment we must quantize the electromagnetic field and introduce specific representations for the field. A brief summary of the main results required for our analysis is given in Sec. II.

II. QUANTIZATION OF THE ELECTROMAGNETIC FIELD

The electromagnetic field may be written as the sum of positive and negative frequency parts

$$E(\underline{r},t) = E^{(+)}(\underline{r},t) + E^{(-)}(\underline{r},t), \tag{2.1}$$

where

$$E^{(-)}(\underline{r},t) = (E^{(+)}(\underline{r},t))^{\dagger}, \tag{2.2}$$

the dagger signifying Hermitian conjugate.

We may expand $E^{(+)}(\underline{r},t)$ in terms of normal modes as

$$E^{(+)}(\underline{r},t) = i\sum_k \left(\frac{\hbar\omega_k}{2}\right)^{1/2} a_k u_k(\underline{r}) e^{-i\omega_k t}, \tag{2.3}$$

where $u_k(\underline{r})$ are the normal mode functions.

The expansion (2.3) has the same form as the classical expansion where the a_k are interpreted as field amplitudes. To quantize the field, the a_k become operators each possessing identical properties to the normal mode operators of the quantized harmonic oscillator. The operator a_k is interpreted as the annihilation operator for a photon in mode k and $a_k^{\dagger}$ is the corresponding creation operator. These operators obey the boson commutation relations

$$[a_k, a_{k'}^{\dagger}] = \delta_{kk'}$$

$$[a_k, a_{k'}] = [a_k^{\dagger}, a_{k'}^{\dagger}] = 0 \tag{2.4}$$

To demonstrate the action of these operators we consider as a specific representation of the electromagnetic field the Fock or number-state representation

$$|\{n_k\}\rangle = |n_1, n_2 \ldots n_i, \ldots n_\infty\rangle \tag{2.5}$$

This representation implies that there are n_i photons in the ith mode. Since each mode is independent, a complete set of state vectors may be written as a simple product of the state vectors for each mode, that is

$$|\{n_k\}> = \prod_{k=1}^{\infty} |n_k> \tag{2.6}$$

These number states form an orthonormal set

$$<n_k|m_{k'}> = \delta_{nm}\delta_{kk'}. \tag{2.7}$$

The meaning of the creation and annihilation operators becomes clear when considering their operation on these states

$$a_k|n_k> = n_k^{1/2}|n_k - 1>$$

$$a_k^\dagger|n_k> = (n_k + 1)^{1/2}|n_k + 1>. \tag{2.8}$$

The operation of the number operator $a_k^\dagger a_k$ is as follows

$$a_k^\dagger a_k|n_k> = n_k|n_k>. \tag{2.9}$$

An alternative representation for the electromagnetic field is the coherent states

$$|\{\alpha_k\}> = \prod_{k=1}^{\infty} |\alpha_k>. \tag{2.10}$$

These coherent states are eigenstates of the annihilation operator

$$a_k|\alpha_k> = \alpha_k|\alpha_k>, \tag{2.11}$$

and may be created from the vacuum state as follows

$$|\alpha> = \exp(\alpha a^\dagger - \alpha^* a)|0> \tag{2.12}$$

They form a complete set and may be expanded in terms of the number of states as follows

$$|\alpha> = e^{-(1/2)|\alpha|^2} \sum_{n=0}^{\infty} \frac{\alpha^n}{(n!)^{1/2}} |n>. \tag{2.13}$$

Thus, the coherent states possess an indefinite number of photons which are Poisson distributed with mean $|\alpha|^2$.

III. INTERFERENCE EXPERIMENTS

The type of interference experiment we shall consider involves the superposition of two light beams to form an interference pattern on a screen. These experiments may use light beams originating from a common source as in Young's pinhole experiment or may involve the interference of independent light beams as for example with two crossed laser beams.

The interference pattern is detected on a screen at position $\underline{r}$ and time t by a suitable photon detector. A quantum theory of photon detection shows that the field intensity measured by an ideal photon detector is proportional to[3]

$$I(\underline{r},t) = \mathrm{Tr}\{\rho E^{(-)}(\underline{r},t)E^{(+)}(\underline{r},t)\}, \tag{3.1}$$

where ρ is the density operator describing the state of the field. This intensity pattern varies as one moves the point of observation up and down the screen thereby changing the phase relation between the two beams. The in-phase components interfere constructively, while the out-of-phase components interfere destructively, thereby producing alternate bright and dark fringes.

We shall now consider two typical interference experiments.

A. Young's Interference Experiment

We first consider an interference experiment of the type performed by Young which consists of light from a monochromatic point-source S incident on a screen possessing two pinholes P_1 and P_2 which are equidistant from S (Fig.1). The pinholes act as secondary monochromatic point sources which are in phase and the beams from them are superimposed on a screen at position $\underline{r}$ and time t. In this region an interference pattern is formed.

To avoid calculating the diffraction pattern for the pinholes we assume their dimensions are of the order of the wavelength of light in which case they effectively act as sources for single modes of spherical radiation in keeping with Huygen's principle. The appropriate mode functions for spherical radiation are

$$u_k(\underline{r}) = [\hat{e}_k/(4\pi R)^{1/2}](e^{ikr}/r); \tag{3.2}$$

where R is the radius of the normalization volume and $\hat{e}_k$ is a unit-polarization vector.

The field detected on the screen at position $\underline{r}$ and time t is then the sum of the two spherical modes emitted by the two pinholes

$$E^{(+)}(\underline{r},t) = f(\underline{r},t)(a_1e^{ikr_1} + a_2e^{ikr_2}), \tag{3.3}$$

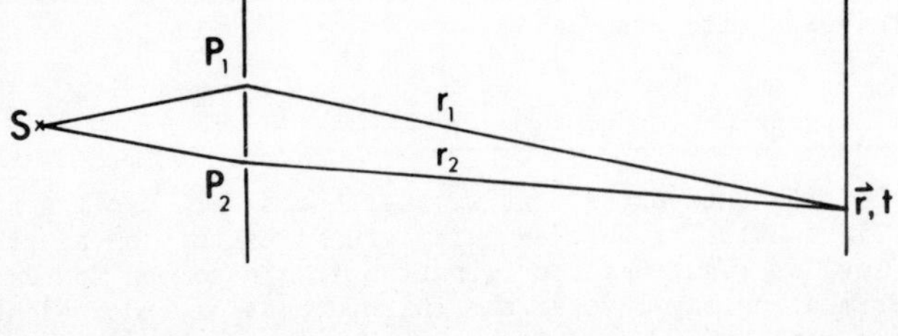

Fig. 1 Young's experiment.

where

$$f(\underline{r},t) = i\left(\frac{\hbar\omega}{2}\right)^{1/2} \frac{\hat{e}_k}{(4\pi R)^{1/2}} \frac{1}{r} e^{-i\omega t}, \tag{3.4}$$

where r_1 and r_2 are the distances of the pinholes P_1 and P_2 to the point on the screen, and we have set $r_1 \approx r_2 = r$ in the denominators of the mode functions.

Substituting Eq.(3.3) into the expression (3.1) for the intensity, we find

$$I(\underline{r},t) = N[\mathrm{Tr}(\rho a_1^\dagger a_1) + T_r(\rho a_2^\dagger a_2) + 2|\mathrm{Tr}(\rho a_1^\dagger a_2)| \cos\Phi], \tag{3.5}$$

where

$$\mathrm{Tr}(\rho a_1^\dagger a_2) = |\mathrm{Tr}(\rho a_1^\dagger a_2)| e^{i\phi},$$

$$N = |f(\underline{r},t)|^2,$$

$$\Phi = k(r_1 - r_2) + \phi. \tag{3.6}$$

This expression exhibits the typical interference fringes with the maximum of intensity occurring at

$$k(r_1 - r_2) + \phi = n2\pi \tag{3.7}$$

with n an integer.

The maximum intensity of the fringes falls off as one moves the point of observation further from the central line by the $1/r^2$ factor in $|f(\underline{r},t)|^2$.

B. Interference of Two Plane Waves

We now consider the interference arising from two plane waves with wave vectors $\underline{k}_1$ and $\underline{k}_2$ intersecting at an angle θ (Fig. 2). Such a situation is approximated by the intersection of two ideal laser beams. Interference fringes may then be observed along the direction $\underline{r}$ perpendicular to $\underline{k}_1$, since the wave $\underline{k}_2$ has a varying phase in this direction.

The appropriate plane-wave mode functions are

$$u_k(\underline{r}) = \hat{e}_k(1/V^{1/2})e^{i\underline{k}\cdot\underline{r}}, \tag{3.8}$$

where V is the normalization volume. We assume that the two beams originate from monochromatic fully polarized sources and that these sources have identical frequencies, that is $|\underline{k}_1| = |\underline{k}_2| = k$. This is the usual case experimentally since normally these experiments are done by splitting a single laser beam then recombining. This ensures a constant phase relation is maintained between the beams. Interference fringes have been observed with independent light beams; however, this requires the use of highly stabilized laser souces[4].

The field strength at a position $\underline{r}$ and time t on the screen is then just the superposition of the two plane-wave modes. Thus, the mode expansion for the electric field Eq.(2.3) reduces to

$$E^{(+)}(\underline{r},t) = f(t)(a_1 + a_2 e^{i(kr\sin\theta + \varphi)}), \tag{3.9}$$

where

$$f(t) = i\left(\frac{\hbar\omega}{2}\right)^{1/2} \frac{1}{V^{1/2}} e^{-i\omega t}, \tag{3.10}$$

where the direction $\underline{r}$ has been chosen orthogonal to $\underline{k}_1$ and we have included a phase difference φ to allow for the posibility of independent sources.

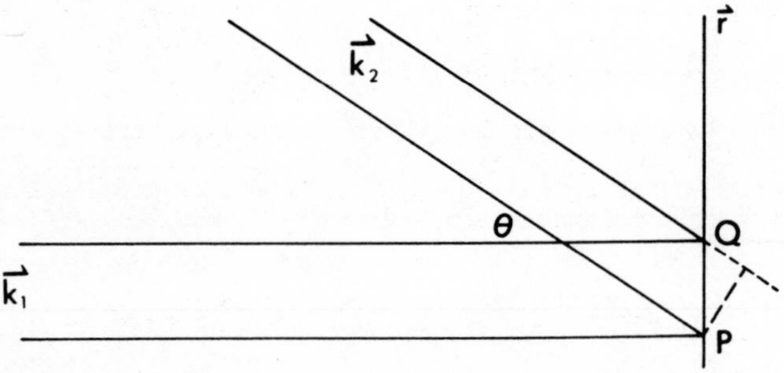

Fig. 2 Interference of two plane waves.

Substituting Eq.(3.10) into Eq.(3.1) we obtain an expression of the form (3.5) for the intensity with

$$N = |f(t)|^2$$

$$\Phi = k\sin\theta r + \phi + \varphi. \tag{3.11}$$

Thus, interference fringes are obtained with constant maximum intensity across the region of the intersecting beams. The maxima of intensity occur at

$$rk\sin\theta = 2n\pi - \phi - \varphi. \tag{3.12}$$

The distance between adjacent fringes d corresponds to a path difference $d\sin\theta$ equal to λ.

Inspection of Eq.(3.5) for the intensity reveals that the essential element for finding interference fringes is that the factor $Tr(\rho a_1^\dagger a_2)$ has a nonzero value. That is, that some degree of correlation exist between the two modes. This is brought out more clearly if we consider the fringe visibility ν which is defined as

$$\nu = (I_{max} - I_{min})/(I_{max} + I_{min}). \tag{3.13}$$

Using Eq.(3.5) this may be expressed as

$$\nu = g[2(I_1 I_2)^{1/2}/(I_1 + I_2)], \quad (3.14)$$

where I_1 and I_2 are the intensities of the two modes, and the normalized correlation function is defined as

$$g = N[|\mathrm{Tr}(\rho a_1^\dagger a_2)|/(I_1 I_2)^{1/2}]. \quad (3.15)$$

This is a specialization of the first-order correlation function of the electromagnetic field to the case where there are only two field-modes present. Utilizing the positive definite nature of the intensity one may prove that[3]

$$N|\mathrm{Tr}(\rho a_1^\dagger a_2)| \lesseqgtr (I_1 I_2)^{1/2}. \quad (3.16)$$

Thus, the values assumed by g may range from zero to one. Fields with g equal to one are considered to possess full first order coherence.[3,5] Fields with $0 < g < 1$ are said to be partially coherent.[6]

It is apparent from Eq.(3.14) that one obtains maximum fringe visibility ($\nu = 1$) from fields which possess first-order optical coherence together with equal intensities in each mode ($I_1 = I_2$). These concepts are demonstrated explicitly in Section IV for several different fields.

IV. INTERFERENCE PATTERNS FOR VARIOUS STATES OF THE RADIATION FIELD

In order to obtain explicit expressions for the intensity obtained in these interference experiments we must evaluate the traces in Eq.(3.5). What follows is common to both the interference experiments discussed in Section III.

We consider the radiation field to be in a pure state with state vector $|\psi\rangle$, that is, possess a density operator given by

$$\rho = |\psi\rangle\langle\psi|. \quad (4.1)$$

We shall begin by noting that fields which possess first-order optical coherence may be shown[7] to result from a single-mode excitation. A general representation of such fields is therefore given by

$$|\psi\rangle = f(b^\dagger)|0\rangle, \quad (4.2)$$

where $|0\rangle$ represents the vacuum state of the radiation field and $b^\dagger$ is the creation operator for a single mode of the radiation field. The operator $b^\dagger$ may be expressed as a linear combination of the creation operators of the two modes involved in the interference experiments discussed in Section III as follows:

$$b^\dagger = a_1^\dagger \cos\theta + a_2^\dagger \sin\theta. \quad (4.3)$$

This operator clearly obeys boson commutation relations

$$[b,b^\dagger] = 1. \tag{4.4}$$

In the following we shall consider several examples of fields generated by this single mode excitation.

A. One Photon Field

A photon field may be generated from a single-mode excitation as follows

$$|1\ \text{photon}\rangle = b^\dagger|0\rangle = \cos\theta|1,0\rangle + \sin\theta|0,1\rangle, \tag{4.5}$$

where the notation used for the eigenkets $|n_1,n_2\rangle$ implies there are n_1 photons present in mode k_1 and n_2 photons present in mode k_2.

Using this state vector to evaluate the traces in Eq.(3.5) for the intensity, the terms which yield a nonzero contribution are

$$\begin{aligned} I(\underline{r},t) = \mathcal{N}\{&\cos^2\theta\ \langle 1,0|a_1^\dagger a_1|1,0\rangle \\ &+ \sin^2\theta\langle 0,1|a_2^\dagger a_2|0,1\rangle \\ &+ \sin 2\theta\ \cos\Phi\langle 1,0|a_1^\dagger a_2|0,1\rangle\}. \end{aligned} \tag{4.6}$$

Using the relations (2.8) together with the orthonormality relations (2.7) this reduces to

$$I(\underline{r},t) = \mathcal{N}(1 + \sin 2\theta\ \cos\Phi). \tag{4.7}$$

It is apparent from this equation that an interference pattern may be built up from a succession of one-photon interference experiments. This one-photon field clearly possesses first-order coherence ($g = 1$) and exhibits maximum fringe visibility for equal intensities in both modes ($\theta = \pi/4$). The above example clearly supports the interpretation that the interference arises from a single photon, as it were, interfering with itself.

This interpretation may not be as readily apparent in the case of two independent light sources, since if the photon is visualized too closely as a particle with classical properties, one might think it had to originate in one or another source. The correct interpretation in this case is nicely summed up by Mandel in his article on photon interference.[8] The Heisenberg uncertainty principle states that the product of uncertainty in detected position Δr (which we take sufficiently large to correspond to a photodetection) and uncertainty in momentum Δk obeys $\Delta k \Delta r > \hbar$. "Thus localization of a photon at $\underline{r},t$ with a resolution better than one fringe width undetermines the momentum of the photon to such an extent that it is impossible to ascribe it to either of the two sources separately."

B. n Photon Field

An obvious generalization of the one-photon field is the n photon field generated by a single-mode excitation. This field is described by the wave function

$$|n\ \text{photon}\rangle = [(b^\dagger)^n/(n!)^{1/2}]|0\rangle$$

$$= \sum_{j=0}^{\infty} (\cos\theta)^j(\sin\theta)^{n-j}\left(\frac{n!}{j!(n-j)!}\right)^{1/2}|j,n-j\rangle. \quad (4.8)$$

For simplicity, we shall treat the two-photon field and assume that the beams have equal intensities $\theta = \pi/4$. The wave function for this two-photon field is

$$|2\ \text{photon}\rangle = \tfrac{1}{2}(|2,0\rangle + \sqrt{2}|1,1\rangle + |0,2\rangle). \quad (4.9)$$

Using this wave function to evaluate the traces in Eq. (3.5) for the intensity yields

$$I(\underline{r},t) = \frac{N}{4}\{\langle 2,0|a_1^\dagger a_1|2,0\rangle + 2\langle 1,1|a_1^\dagger a_1|1,1\rangle$$

$$+ \langle 0,2|a_2^\dagger a_2|0,2\rangle + 2\langle 1,1|a_2^\dagger a_2|1,1\rangle$$

$$+ 4\sqrt{2}\cos\Phi\langle 0,2|a_2^\dagger a_1|1,1\rangle\}$$

$$= 2N(1 + \cos\Phi), \quad (4.10)$$

which as expected exhibits maximum fringe visibility.

Note that it is possible to construct, mathematically at least, a two-photon field which gives rise to no interference fringes whatsoever. An example of such a field would be the state with wave function

$$|2\ \text{photon}\rangle' = (1/\sqrt{2})(|2,0\rangle + |0.2\rangle). \quad (4.11)$$

That is, there is an equal probability of there being two photons in either mode, but zero probability for there being one photon in each mode. Such a field cannot be generated by a single-mode excitation. This field clearly has a zero correlation function g and hence, exhibits no interference fringes. Comparison of Eqs.(4.11) and Eq.(4.9) show that it is necessary to have states differing in occupation by a single photon in order to obtain a nonzero value for the correlation function.

C. Coherent Field

As a final example we consider the case where both modes are in coherent states of the radiation field. As mentioned in Section III, a coherent field is characterized by a Poissonian photon number distribution. Such a field may be produced by an ideal laser. The wave function for this coherent field is

$$|\text{coherent field}\rangle = |\alpha_1,\alpha_2\rangle = |\alpha_1\rangle|\alpha_2\rangle. \quad (4.12)$$

Since this wave function is a product state it may well represent two independent light beams. This particular product state may however be generated by a single-mode excitation in the following manner

$$|\alpha_1\rangle|\alpha_2\rangle = \exp(ab^\dagger + a^*b)|0\rangle$$

$$= \exp(\alpha a_1^\dagger \cos\theta + \alpha^* a_1 \cos\theta)$$

$$\times \exp(\alpha a_2^\dagger \sin\theta + \alpha^* a_2 \sin\theta)|0\rangle$$

$$= |\alpha \cos\theta\rangle|\alpha \sin\theta\rangle., \tag{4.13}$$

which follows using Eq.(2.12).

With the help of Eq.(2.11) and its adjoint to evaluate the traces in Eq.(3.5), the intensity pattern produced by this coherent field is found to be

$$I(\underline{r},t) = \mathcal{N}(|\alpha|^2 + |\alpha|^2 \sin 2\theta \cos\Phi), \tag{4.14}$$

which for equal intensities in both modes exhibits maximum fringe visibility. Thus, the coherent states possess first-order optical coherence (in fact, as their name implies, they possess coherence to all orders[3,5]).

The above example demonstrates the possibility of observing interference between independent light beams. Experimentally, this requires that the phase relation between the two beams be slowly varying or else the fringe pattern will be washed out. Interference between independent light beams is, however, only possible for certain states of the radiation field, for example, the coherent states as demonstrated above. Interference is not generally obtained from independent light beams as we demonstrate in the following example. We consider the two independent light beams to be in Fock states, that is, described by the wave function

$$|\psi\rangle = |n_1\rangle|n_2\rangle. \tag{4.15}$$

This yields a zero correlation function and consequently no fringes are obtained.

The above demonstration of interference arising from the coherent states obscures somewhat the underlying one photon mechanism. If, however, we recall that the coherent states may be expressed as a sum of number states [Eq.(2.13)] we see that the interference arises between the states differing in occupation number by one photon. These are phased in such a way as to give maximum correlation (g = 1).

V. SUMMARY

We have presented a simple field-theoretic description of an optical interference experiment of the Young's type. The formalism used allows one to calculate the interference pattern for any arbitrary quantum state of the radiation field. In particular, interference fringes were shown to arise from a field containing only one photon, thus giving an explicit example of Dirac's statement, "Each photon then interferes only with itself." The interference patterns produced by other photon fields including an n photon and a coherent photon field were also investigated.

ACKNOWLEDGMENTS

The author is particularly grateful to Mr. H.J. Carmichael for helpful comments on the manuscript.

The author is grateful for comments by the referees which have contributed to the final form of this paper.

REFERENCES

1. Born M and Wolf E *Principles of Optics*, 3rd ed. (Pergamon, New York, 1965)
2. Dirac P A M *The Principles of Quantum Mechanics*, 3rd ed. (Clarendon Press, Oxford, 1947), p.9.
3. Glauber R J in *Quantum Optics and Electronics*, edited by C. de Witt, A. Blandin and C. Cohen Tannoudji (Gordon and Breach, New York, 1964).
4. Pfleegor R L and Mandel L 1967 *Phys.Rev. 159* 1084; 1968 *J. Opt. Soc. Am. 58* 946
5. Klauder J R and Sudarshan E C G *Fundamentals of Quantum Optics* (Benjamin, New York, 1968)
6. A comprehensive review article on partial coherence has been written by Sudarshan E C G 1969 *J. Math. Phys. Soc. 3* 121
7. Titulaer U M and Glauber R J 1965 *Phys. Rev. 140* B676; 1966 *Phys. Rev. 145* 1041
8. Mandel L *Quantum Optics* edited by R.L. Glauber (Academic, New York, 1969), p.182